ACE YOUR MIDTERMS & FINALS

U.S. HISTORY

Other books in the Ace Your Midterms and Finals Series include:

Ace Your Midterms and Finals: Introduction to Psychology

Ace Your Midterms and Finals: Principles of Economics

Ace Your Midterms and Finals: Fundamentals of Mathematics

Ace Your Midterms and Finals: Introduction to Physics

Ace Your Midterms and Finals: Introduction to Biology

ACE YOUR MIDTERMS & FINALS

U.S. HISTORY

ALAN AXELROD, PH.D.

McGraw-Hill

New York San Francisco Washington, D.C. Auckland Bogotá
Caracas Lisbon London Madrid Mexico City Milan
Montreal New Delhi San Juan Singapore
Sydney Tokyo Toronto

Library of Congress Cataloging-in-Publication Data

Axelrod, Alan
Ace your midterms and finals: U.S. history / Alan Axelrod.
p. cm. — (Ace your midterms and finals series)
Includes index.
ISBN 0-07-007005-9
1. United States—History—Examinations—Study guides. 2. United States—History Examinations, questions, etc. I. Title.
II. Title: U.S. History. III. Series: Axelrod, Alan, 1952- Ace your midterms and finals series.
E178.25.A94 1999
973'.076—dc21 99-24818
CIP

McGraw-Hill
A Division of The ***McGraw-Hill*** *Companies*

1 2 3 4 5 6 7 8 9 0 DOC/DOC 9 0 4 3 2 1 0 9

ISBN 0-07-007005-9

The sponsoring editor for this book was Barbara Gilson, the editing supervisor was Frank Kotowski, Jr., and the production supervisor was Modestine Cameron. It was set in Minion by Carol Norton for The Ian Samuel Group, Inc.

Printed and bound by R. R. Donnelley & Sons Company.

This book is printed on recycled, acid-free paper containing a minimum of 50% recycled, de-inked fiber.

CONTENTS

PART THREE
MIDTERMS AND FINALS

PART FOUR
FOR YOUR REFERENCE

HOW TO USE THIS BOOK

YOU KNOW THE DRILL. FIRST DAY IN A SURVEY, INTRODUCTORY, OR "CORE COURSE: the professor talks about grading and, saying something about the value of the course in a program of "liberal education," declares that what she wants from her students is original thought and creativity and, above all, that she does not "teach for" the midterm and final.

Nevertheless, the course certainly *includes* one or two midterms and a final, which account for a very large part of the course grade. Maybe the professor can disclaim with a straight face *teaching for* these exams, but few students would deny *learning for* them.

True, you know that the purpose of an introductory course is to gain a useful familiarity with a certain field and *not* just to prepare for and do well on a couple or three exams. Yet the exams *are* a big part of the course, and, whatever you learn or fail to learn in the course, your performance as a whole is judged in large measure by your performance on these exams.

So the cold truth is this: More than anything else, curriculum core courses *are* focused on the midterm and final exams.

Now, traditional study guides are outlines that attempt a bird's-eye view of a given course. But *Ace Your Midterms and Finals: U.S. History* breaks with this tradition by viewing course content through the magnifying lens of ultimate accountability: the course exams. The heart and soul of this book consists of eleven midterms and eleven finals prepared by *real* instructors, teaching assistants, and professors for *real* students in *real* schools.

Where did we get these exams? Straight from the professors and instructors themselves.

- ◆ All exams are real and have been used in real courses.
- ◆ All exams include critical "how-to" tips and advice from the creators and graders of the exams.
- ◆ All exams include actual answers.

Let's talk about those answers for a minute. In most cases, the answers are actual student responses to the exam. In some cases, however, the instructors and professors have created "model" or "ideal" answers. Usually, the answers included are A-level responses. Sometimes, however, they are not perfect (because they are real). In all cases, you'll find full commentary that points out what works (and why) and what could use improvement (and why—as well as how to improve it).

This book also contains more than the exams themselves.

- ◆ In "Part One: Preparing Yourself," you'll find how-to guidance on what history professors look for, how to think like a historian, how to study more effectively, and how to gain the performance edge when you take an exam.
- ◆ "Part Two: Study Guide" presents a quick-and-easy overview of the content of a typical survey of American history. It clues you in on what to expect in the course.
- ◆ "Part Three: Midterms and Finals" are the exams themselves, grouped by college or university.
- ◆ In "Part Four: For Your Reference," you'll find a handy chronology of the major events in U.S. history and a list of recommended reading, arranged by topic area.

What This Book *Is Not*

Ace Your Midterms and Finals: U.S. History offers a lot of help to see you through to success in this important course. But (as you'll discover when you read Part One) the book *cannot* take the place of

- ◆ Doing the assigned reading

- Keeping up with your work and study
- Attending class
- Taking good lecture notes
- Thinking about and discussing the topics and issues raised in class and in your books

Ace Your Midterms and Finals: U.S. History is not a substitute for the course itself!

What This Book *Is*

Look, it's both cynical and silly to invest your time, brainpower, and money in a college course just so that you can ace a couple of exams. If you get A's on the midterm and final, but come away from the course having learned nothing, you've failed.

We don't want you to be cynical or silly. The purpose of introductory, survey, or "core" courses is to give you a panoramic view of the knowledge landscape of a particular field. The primary goal of the college experience is to acquire more than tunnel intelligence. It is to enable you to approach whatever field or profession or work you decide to specialize in from the richest, broadest perspective possible. College is education, not just vocational training.

We don't want you to "study for the exam." The idea is to study for "the rest of your life." You are buying knowledge with your time, your brains, and your money. It's an expensive and valuable commodity. Don't leave it behind you in the classroom at the end of the semester. Take it with you.

But even the starriest-eyed idealist can't deny that midterms and finals are a big part of intro courses and that even if your ambitions lie well beyond these exams (which they should!), performing well on them is necessary to realize those loftier ambitions.

Don't, however, think of midterms and finals as hurdles—obstacles—you must clear in order to realize your ambitions and attain your goals. The exams are there. They're real. They're facts of college life. You might as well make the most of them.

Use the exams to help you focus your study more effectively. Most people make the mistake of confusing *goals* with *objectives.* Goals are the big targets, the ultimate prizes in life. Objectives are the smaller, intermediate steps that have to be taken to reach those goals.

Success on midterms and finals is an objective. It is an important, sometimes intimidating, but really quite *doable* step toward achieving your goals. Studying for—working toward—the midterm or final is *not* a bad thing, as long as you keep in mind the difference between objectives and goals. In fact, fixing your eye on the upcoming exam will help you to study more effectively. It gives you a more urgent purpose. It also gives you something specific to set your sights on.

And this book will help you study for your exams more effectively. By letting you see how knowledge may be applied—immediately and directly—to exams, it will help you acquire that knowledge more quickly, thoroughly, and certainly. Studying these exams will help you to focus your study in order to achieve success on the exams—that is, to help you attain the objectives that build toward your goals.

—Alan Axelrod

CONTRIBUTORS

Jason M. Barrett, *Instructor, University of Michigan (Chap. 21)*

Kathryn Fenn, *Doctoral Candidate, Duke University (Chap. 17)*

Kevin Gannon, *Graduate Instructional Assistant, University of South Carolina (Chap. 24)*

Colin Gordon, *Associate Professor, University of Iowa (Chap. 20)*

Aram Goudsouzian, *Graduate Teaching Assistant, Purdue University (Chap. 23)*

Timothy P. Maga, *Oglesby Professor of American Heritage, Bradley University (Chap. 15)*

Daniel S. Murphree, *Graduate Teaching Assistant, Florida State University (Chap. 19)*

Wendy St. Jean, *Lecturer, University of Connecticut (Chap. 16)*

Margaret Storey, *Teaching Assistant, Emory University (Chap. 18)*

Richard M. Ugland, *Academic Program Coordinator/Lecturer, Ohio State University (Chap. 22)*

Stephanie E. Yuhl, *Post-doctoral Fellow, Valparaiso University (Chap. 25)*

ABOUT THE AUTHOR

Alan Axelrod, Ph.D., is the author of many books, including *Booklist* Editor's Choice *Art of the Golden West, The Penguin Dictionary of American Folklore,* and *The Macmillan Dictionary of Military Biography*. He lives in Atlanta, Georgia.

PART ONE

PREPARING YOURSELF

CHAPTER 1

THE U.S. HISTORY SURVEY: WHAT THE PROFESSORS LOOK FOR

AMERICAN HISTORY: DATES AND EVENTS, RIGHT? FROM COLUMBUS IN 1492 TO WHATever happened when the course ends, right?

Not really. There is a big difference between chronology—dates and events—and history, which is the study of the past. The key word here is *study*. Chronology is a list. It can be memorized. (It's no fun, but it can be done.) But history, as it is taught in most colleges and universities, cannot simply be committed to memory. It is as much about the relationships between events and between people as it is about dates and summary facts. The *study* of history, then, is not primarily a matter of memory—although memory is important—but of analysis and critical thinking.

Analytical thinking . . . is more important than rote memory efforts or guess work. In today's global and highly competitive marketplace, the history student cannot afford to be lacking in analytical prowess. Examining historical dilemmas and controversies, many of which are relevant to the present, assists in building that prowess.

—Timothy P. Maga, Bradley University

Analysis requires gathering and examining the facts of history and drawing conclusions about the relationships among them—especially cause-and-effect relationships—and their significance. You would be quite correct in recalling, for example, that the U.S. naval base at Pearl Harbor was attacked by the Japanese on December 7, 1941. But, unless you are answering a multiple-choice or fill-in-the-blank test question, this response will do little to impress a history professor. You would need to go on to talk about the significance of the event—primarily how it propelled the United States into World War II. Further analysis might go on to consider just why it took such a devastating attack to shake the United States out of neutrality, and it might also cover how the attack, far from demoralizing or intimidating Americans, unified them in purpose and resolve.

Thus history professors expect you two attain two broad goals:

1. To master a body of historical knowledge—the events and dates, the "facts."
2. To analyze the relationships among those facts; to draw conclusions about the *significance* of the facts.

Don't worry about the details. It is more important to have a grasp of the basic issues and of how they change over time.

—Colin Gordon, University of Iowa

Reaching the first goal requires reading assigned texts and attending classes and lectures. Reaching the second goal requires the following:

1. Undertaking reading and research projects in addition to textbook assignments.
2. Developing research skills.
3. Developing critical reading skills.
4. Developing analytical thinking skills.
5. Developing discussion and oral presentation skills.
6. Developing analytical writing skills.

You can see from this list that although your professor certainly expects you to read assigned material and to pay attention in class, she expects even more that you will not simply "absorb" material, but develop at least five important skills. That history requires more than hitting the books and taking lecture notes comes as a shock to some students. They expect to be required to develop special "skills" in art classes, say, or in physical education and sports electives. But in *history*?

Well, take a good look at that list of skills. Some students grumble that history is a bunch of facts about dead people: "I'm going to be a corporate lawyer. Why do I need to memorize Woodrow Wilson's 'Fourteen Points'?" We could argue with such a grumbler that learning the facts of history helps us understand who we are as Americans. And we could also repeat what the philosopher George Santayana famously said: "Those who do not know history are condemned to relive it." But, instead, let's concede the point. It is true: familiarity with the international policies of a long-dead pres-

Do NOT sit and memorize all the facts and dates you can. Successful students do not worry about facts and dates. They try to integrate the material they already have into coherent wholes. They come up with explanations of relationships between issues. Come into the exam with a few ideas that explain the relationship between economics and religion, or the problems of political ideology, or race and ethnicity. Odds are, you can make at least one of your ideas fit into one of the questions offered.

—Jason M. Barrett,
University of Michigan

ident may not someday help you to negotiate a corporate contract. Fortunately, an introductory history course presents an opportunity to do much more than memorize facts about the dead. Success in the course requires developing and honing that list of skills—and the application of these skills is hardly limited to the academic study of history. What lawyer (or corporate executive, or salesperson, or teacher, or physician) could fail to benefit from an ability to find information quickly and efficiently (develop research skills), to understand and evaluate information (develop critical reading skills), to do something productive with information gathered (develop analytical thinking skills), to present information clearly, persuasively, and effectively (develop discussion and oral presentation skills and analytical writing skills)?

Question:
What key competencies, skills, quality of thought, and/or specific kinds of information are you looking for in students' responses to examinations?

Answer:

1. **mastery of the course material**
2. **critical and original thinking about the course material**
3. **clarity of expression**

—Jason M. Barrett,
University of Michigan

In short, most of what your history professor expects from you is to develop the skills of a good student—not a passive absorber of information, but an active, thinking *student* of information. These skills will help you in all of your course work, and, even more important, are essential to any reasonably high-level job in the "real world" beyond college.

Engaged actively, critically, and analytically, the facts of history come alive and are far more interesting than any novel or movie for the simple reason that they are real, they affect our lives, and they shape the destiny of our nation and our world. That's great news. However, history courses do present certain pitfalls.

The most common trap into which students fall is mistaking opinion for analysis. You will find your professor repeatedly asking you "what you think" about a certain event or set of data. For example, "Was the United States justified in dropping atomic bombs on Japan during World War II? What do you think?"

Perhaps you have a strong opinion on this topic. Perhaps you have just read John Hershey's *Hiroshima* and have learned something of the horrors wrought by a nuclear detonation. So you think the United States shouldn't have dropped the bombs.

Or perhaps you've just read an account of the way in which Japanese soldiers fought to the death rather than surrender. Or perhaps you've encountered an account of the wholesale Japanese torture of Chinese civilians during the war. Perhaps such information inclines you to approve of the use of nuclear weapons against such an enemy.

On the other hand, perhaps you just *feel* that dropping the bomb was the wrong or the right thing to do.

However you have formed your opinion, you are certainly entitled to it. But will expressing that opinion earn you points in a university history class?

Probably not—unless you transform that *opinion* into *analysis.* This requires supporting your opinion with facts, and not just one or two facts casually encountered (that Hershey book or a single article on the treatment of Chinese civilians), but a whole *body* of relevant facts. For example, if you believe that the United States was justified in using the atomic bomb in World War II, you might bring into evidence casualty figures for the Pacific island campaigns, you might point out projections as to the cost of both Allied and Japanese lives if an invasion of Japan were attempted, you might point out the moral and ethical pressure on President Truman to use all means at his disposal to end this costliest of wars quickly and with whatever means were at hand, no matter how terrible.

Many students in discussion sections of introductory history courses become frustrated because the instructor repeatedly asks "what do you think?" only to shoot down an opinion because it lacks supporting facts. You can avoid this frustration by understanding that when you are asked "what you think," you are being asked for an interpretation of fact rather than a mere opinion. Interpretation of fact involves developing the skills listed a moment ago.

Let's consider those skills briefly.

We live in an age flooded with information. Television, movies, books, and the Internet all yield a continual flow. Not only the best students, but the most powerful and influential people these days are the ones who know how to snatch the most useful bits of information out of the deluge. Research skills involve knowing where to look for information, as well as determining what kinds of information to look for. Your professor expects you to develop such skills, beginning with basic library skills and progressing through the intelligent use of bibliographies, reader's guides, indexes, and even manuscript material.

Students should . . . develop a more complex view of the different issues relevant to U.S. history and be able to ascertain their own personal relationships to these issues. Perhaps most important, students should refine their skills in creating a logical argument and substantiating it with valid evidence.

. . . All students are verbally instructed during the first week of classes on how best to prepare for examinations and research papers. Besides standard reminders concerning the taking of class lecture notes and scheduling of text readings, students are prompted to approach all assignments by considering at least two arguments. When analyzing issues such as U.S. policy toward Native Americans, or public opinion toward the Civil War, students should view the topic from all major perspectives before forming their own personal opinion. The goal is to convince students that much of what they learn about history is interpretation, and much like history professionals, students themselves should develop an interpretation based on all available perspectives and evidence.

—Daniel S. Murphree,
Florida State University

Perhaps you've seen one of those demonstrations common in psych courses, prelaw classes, or introductory criminal justice courses. The instructor is lecturing as usual at the front of the class. Suddenly, someone rushes into the classroom, does and says something outrageous, then promptly exits. After the tumult dies down, the professor turns to the class and asks the students to jot down answers to a set of questions about what they had just witnessed: What was the "intruder" wearing? What was the intruder's gender? How tall? Fat? Thin? What, exactly, did the intruder say? And so on. Invariably, these witnesses will return an array of different and often contradictory answers to the same set of questions. If there are thirty witnesses, there are bound to be thirty quite distinct responses. One person will express certainty that the intruder was a man. Another that she was a woman—dressed in a red sweater. No, says another, it was a pink blouse. Actually, a bright red scarf with a white shirt. And so on.

Course objectives—students should:

(1) Know and be able to identify certain people, places and events significant to United States history.

(2) Analyze and form their own opinions on past trends and movements.

(3) Evaluate the similarities and disparities between North America's various regions and peoples during the development of the United States.

In addition to the stated course objectives, students should improve both their research and writing skills and broaden their ability to interpret different materials through a relatively objective perspective.

—Daniel S. Murphree, Florida State University

Books, diaries, manuscripts, personal accounts, all the written records of history are human perspectives on events. For many and varied reasons, these perspectives will differ, even when they focus on the same set of events—just as eyewitness descriptions will vary from person to person. Your instructor will expect you to develop some skill in reading the materials of history critically. If the question is "Was the Civil War fought over the issue of slavery?" and you limit your supporting readings to the extreme pronouncements of South Carolina's John Calhoun, you'll answer no. If, however, you focus exclusively on the writings of radical northern abolitionists, you'll emerge with an unequivocal yes. And if you look at what Abraham Lincoln said on the subject, you'll come out somewhere in the middle. Critical reading requires weighing the arguments made by authors, understanding their motives, and knowing "where they're coming from." Your instructor expects you to question what you read, not to take it at face value.

We've all had the discouraging experience of plowing through a book, reading every word, closing the cover, and then realizing that we have comprehended absolutely none of it. Finishing all reading assignments in a timely manner is important, but doing so is of little value if you fail to think about what you've read.

Emphasis is on learning how to read both primary and secondary sources critically. Students should become sensitized to authors' biases and to writers' use of evidence to support their cases.

—Wendy St. Jean, University of Connecticut

Now, the phrase "think about" is pretty vague and abstract. Let's be more specific about what to do with what you've read. *Connect* one fact, one event, to another. Maybe you know the dates for the Battle of Gettysburg. Maybe you know the names of the chief commanders in that battle. Now, why was it fought? What happened as a result of the Union victory? What did the battle lead to? What might have happened had the South prevailed in the battle? What would that have led to?

Reading critically and thinking analytically are important skills, but they're pretty lonely. The products of critical analysis are intended to be shared. Besides, until you communicate your observations, thoughts, and conclusions, your instructor has no way to evaluate your performance and progress in the course. For this reason, your instructor expects you to develop skills in discussing issues and presenting arguments.

Tips for critically reading a history book

Ask:

1. **Who wrote this?**
2. **When was this written?**
3. **Is there an attempt to present a balanced argument, or does the author have an axe to grind?**
4. **What sources does this author use?**
5. **To what additional sources does this author send me?**

Class discussion is a balancing act. It is, unfortunately, the rare class or discussion section in which all students participate. Typically, the majority hang back and take notes while the instructor chats with a vocal minority. Your instructor expects you to participate in class discussion. Usually, the more you participate, the better. But let's pay attention to the word *usually*. It's all too easy to fall in love with the sound of your own voice. Don't monopolize the discussion. Listen to the interpretations of others. Also, practice courtesy. Don't patronize, mock, or put anyone down for a remark that is made. Learn to focus your arguments on other arguments, not on personalities. Instead of saying, "Joe, you're just wrong about the New Deal," ask a question about the subject: "Well, Joe, what about the cost of the New Deal programs?" The object of class discussion is not to win an argument, but to explore all sides of an issue. Keep the issues in the foreground, and personalities out of the picture completely.

The main skill is communication—both written and oral. [I want to help] students to develop critical thinking and effective communication skills while deploying historical data from the American Experience. Students are also encouraged to be broad-minded and sensitive in their approach to the variety of ways in which people experienced the past. They are also encouraged to respect each other's ideas and claims, while nonetheless feeling free to challenge each other, as long as they have data to back up their opposing claims.

—Stephanie E. Yuhl, Valparaiso

Writing comes hard for many students, and the very phrase "analytical writing" can be intimidating. Good analytical writing requires thorough familiarity with your subject, a willingness to find and use the right words, and an ability to organize your ideas. These three elements require dedication and hard work; however, there are some shortcuts you should take advantage of.

First, when you are responding to essay questions in an exam or other assignment, look at the question carefully. What exactly does it ask? Is it made up of a number of component questions—subquestions? In what order do these occur? Look at the question, and try to use it to structure your essay response. That is, if the question asks you to discuss points 1, 2, 3, and 4, structure your essay in that order. If it asks for 2, 4, 3, and 1, try to respond accordingly, answering 2, 4, 3, and 1 rather than 1, 2, 3, and 4.

Second, most writing about history is naturally chronological. Unless you have a very good reason to do otherwise, organize your writing chronologically, following the order of events.

While it is true that introductory history courses challenge you to develop a number of advanced skills, you will find that extra time and effort devoted to honing these skills is time very well invested. Not only can you expect a high level of performance in history class, but you will find these skills of thought and expression basic to just about every other course you take.

With business and government hiring more history majors and minors than ever before, it is essential, I believe, that students have a decent background in analytical writing. Their ability to argue a position clearly is important to on-the-job success after graduation. Hence, my stress on analytical essay exams versus the multiple guess approach or even the short-answer format. For too long, the history course has been associated with the multiple choice/true-or-false exam, thereby keeping the student away from analytical thinking and more involved in rote memory efforts.

—Timothy P. Maga, Bradley University

So far, we have discussed some general expectations your history instructor is likely to have. Let's look at some important expectations that are more specific to the subject of history.

You will probably hear a good deal of talk about the "historical method." The historical method is more a habit of thought than a specific method. It's learning ways to help distinguish truth from distortion and outright falsehood. It's practicing how to separate opinion from interpretation based on fact. It's a habit of looking at events in terms of cause and effect.

You will be exposed to a variety of historical sources, ranging from primary sources—the equivalent of eyewitness accounts (just remember how unreliable eyewitnesses can be!)—to secondary sources (mainly books and articles written after-the-facts by historians). You will have to become accustomed to assessing and evaluating your sources, so that you can judge their reliability.

And let's not forget facts. The one thing most history instructors do *not* want is a classroom full of students parroting names and dates. So they tend to deemphasize facts and rote learning. Yet the one *fact* that the history student cannot escape is that *facts* are the elements, the building blocks, of history. You will find it easier to comprehend and retain the facts of history if, as you read, you ask yourself the five *w*'s:

Who?
Where?
When?
What?
Why?

When you encounter a name of a person (who), put him in his place (where) and time (when). Associate him with some action, statement, or position (what), and be prepared to answer *why* he did or said or believed whatever he did, said, or believed. This is the kind of five-dimensional thinking your instructor expects. It takes work, but it also makes history come alive. People, after all, aren't flat figures. They're living, breathing beings: multi-dimensional.

There is something else your instructor expects you to do with the information you gather: *make it your own.* Many people read themselves to sleep at night. Curl up with a good book, read a few pages, nod off. Well, reading is one of life's great pleasures, and it can be most soothing. The object of reading your history texts, however, is not sleep. Get out of the habit of reading passively, of the soothing (all *too* soothing) habit of merely taking in words. Instead, learn to read aggressively. Take notes. Underline important passages. Summarize each chapter in your own words.

Tips for Evaluating Primary Sources
Ask:
1. Who wrote this?
2. What makes him/her tick? Prejudices? Biases? Hidden agendas?
3. When was this written?
4. What was the purpose of the document?
5. Who was the audience for it?
6. What does it say? (Take notes.)
7. How does this document fit with what I know?
8. What else do I need to read?

In your own words. That's important. One of the most effective ways of learning is to rephrase material in your own words. Take the words in—then revolve them around in your mind and imagination, so that you can express them in language you are comfortable with and that is most meaningful to you.

Finally, don't rely exclusively on words. The language of history encompasses maps, charts, graphs, diagrams, and other illustrations in addition to words. Take the time to study and understand any maps that you find in your textbook. Geography is central to history, which is, after all, the story of trade patterns, borders, immigration, climate, and invasion. Geography and the tools of geography are important to understanding such events and movements. Population figures, economic statistics, and the like are also important clues to historical reality. When you encounter such data, think about their significance. Draw conclusions from geography and from the numbers associated with geography, economics, and history.

The study of history is challenging, and, make no mistake, you are expected to master a good many facts. But, even more important, you are expected to make those facts sufficiently your own so that you can use them to construct plausible historical interpretations. More than in most introductory-level academic subjects, you are required to exercise judgment, especially in evaluating motives and biases. And to a greater degree than in many other courses, you are called on to present facts and the interpretation of facts in an orderly and persuasive manner.

Does all this sound a bit overwhelming?

Well, there is an upside to an introductory American history course, too. While the study of history calls for many skills, the subject also appeals to interests that probably come naturally to you. Whatever else history consists of, it is a collection of many stories—and most of us enjoy a good story. History also appeals to natural curiosity and the desire most of us have to discover something of our collective past—where we, as a nation, come from. While you develop the challenging skills the study of history requires, be sure not to let go of the many pleasures history offers.

CHAPTER 2

THINKING LIKE A HISTORIAN: KEYS TO SUCCESSFUL STUDY

THIS CHAPTER OUTLINES SKILLS THAT ARE INDISPENSABLE TO SUCCESSFUL STUDY, with special emphasis on skills important to the study of history.

Before you can "think like a historian," you need to *think*. Period. Let's start thinking. After all, isn't that what college and college courses are all about?

Well, not quite. *Think* about it.

For the ancient Greeks of Plato's day, about 428 to 348 B.C., "higher education" really was all about thinking. Through dialogue, back and forth, the teacher and his student *thought* about history, mathematics, physics, the nature of reality—whatever. Perhaps the teacher evaluated the quality of his student's thought, but there is no evidence that Plato graded exams, let alone assigned the student a final grade for the course.

Times have changed.

"Don't Study for the Grade!"

Today, you get graded. All the time, and on everything you do.

Now, most of the professors, instructors, and teaching assistants from whom you take your courses will tell you that the "real value" of the course is in its contribution to your "liberal education." She may even solemnly protest that she does *not* "teach for" the midterm and final. Nevertheless, the introductory, survey, and core courses almost always *include* at least one midterm and a final, and these almost always account for a very large part of the course grade. Even if the professor can disclaim with a straight face *teaching for* these exams, few students would deny *learning for* them.

The truth is this: More than anything else, most curriculum core courses *are* focused on the midterm and final exams.

"Grades Aren't Important"

Let's keep thinking.

You and, most likely, your family are investing a great deal of time and cash and sweat in your college education. It *would* be pretty silly if the payoff of all those resources were a letter grade and a numerical GPA. Ultimately, of course, the payoff is knowledge, a feeling of achievement, an intellectually and spiritually enriched life—*and* preparation for a satisfying and (you probably hope) financially rewarding career.

But the fact is that if you don't perform well on midterms and finals, your path to all these forms of enrichment will be blocked. And the fact *also* is that your performance is measured by grades. Sure, almost any reason you can think of for investing in college is more important than amassing a collection of A's and B's, but those stupid little letters are part of what it takes to get you to those other, far more important, goals.

"Don't Study *for* the Exam"

Most professors hate exams and hate grades. They believe that making students pass tests and then evaluating their performance with a number or letter makes the whole process of education seem pretty trivial. Those professors who tell you that they "don't teach for the exam" may also advise you not to "study for the exam." That's not exactly what they mean. They want you to study, but to study in order to learn, not *for* the exam.

It's well-meaning advice, and it's true that if you study for the exam, intending to ace it, then promptly forget everything you've "learned," you are making a pretty bad mistake. Yet those same professors are part of a system that demands exams and grades, and if you don't study *for* the exam, the chances are very good that you won't make the grade and you won't achieve the "higher goals" you, your family, *and your professors* want for you.

Lose-Lose or Win-Win? Your Choice

When it comes to studying, especially in your introductory-level courses, you have some choices to make. You can decide grades are stupid, not study, and perform poorly on the exams. You can try not to study for the exams, but concentrate on "higher goals," perform poorly on the exams, and never have the opportunity to reach those "higher goals." You can study *for* the exams, ace them all, then flush the information from your memory banks, collect you're A or B for the course, and move on without having learned a thing. These are all lose-lose scenarios, in which no one—neither you nor your teacher (nor your family, for that matter)— gets what he, she, or they really want.

Or you can go the win-win route.

We've used the words *goal* and *goals* several times. For an army general, winning the war is the goal, but to achieve this goal he must first accomplish certain *objectives*, such as winning battle number one, number two, number three, and so on. Objectives are intermediate goals or steps toward an ultimate goal.

Now, put exams in perspective. Performing well on an exam need not be an alternative to achieving higher goals, but should be an objective necessary to achieving those higher goals.

The win-win scenario goes like this: Use the fact of the exams as a way of focusing your study for the course. Focus on the exams as immediate objectives, crucial to achieving your ultimate goals. *Do* study for the exam, but not *for* the exam. *Don't* mistake the battle for the war, the objective for the goal; but *do* realize that you must attain the objectives in order to achieve the goal.

And *that*, ladies and gentlemen, is the purpose of this book:

- ◆ To help you ace the midterm and final exams in introductory-level American history . . .
- ◆ . . . without forgetting everything you learned after you've aced them.

This guide will help you use the exams to master the course material. This guide will help you make the grade—*and* actually learn something in the process.

Focus

How many times have you read a book, word for word, finished it, and closed the covers—only to realize that you've learned almost nothing from it? Unfortunately, it's something we all experience. It's not that the material is too difficult or that its over our heads. It's that we mistake reading for studying.

For many of us, reading is a passive process. We scan page after page, the words go in, and, alas, the words seem to go out. The time, of course, goes by. We've *read* the book, but we've *retained* all too little.

Studying certainly involves reading, but reading and studying aren't one and the same activity. Or we might put it this way: studying is intensely focused reading.

How do you *focus* reading?

Begin by setting objectives. Now, saying that your objective in reading a certain number of chapters in a textbook or reading your lecture notes is to "learn the material" is not very useful. It is an obvious but vague *goal* rather than a well-defined objective.

Why not let the approaching exam determine your objective?

"I will read and retain the stuff in chapters ten through twenty because that's what's going to be on the exam, and I want to ace the exam."

Now you at least have an objective. Accomplish this *objective*, and you will be on your way to achieving the *goal* of "learning the material."

◆ An *objective* is an immediate target. A *goal* is for the long term.

Concentrate

To move from passive reading to active study requires, first of all, concentration. Setting up objectives (immediate targets) rather than looking toward goals (long-term targets) makes it much easier for you to concentrate. Few of us can (or would) put our personal lives "on hold" for four years of college, several years of graduate school, and X years in the working world in order to concentrate on achieving a *goal*. But just about anyone can discipline himself or herself to set aside distractions for the time it takes to achieve the *objective* of studying for an exam.

Step 1. Find a quiet place to work.

Step 2. Clear your mind. Push everything aside for the few hours you spend each day studying.

Step 3. Don't daydream—now. Daydreaming, letting your imagination wander, is actually essential to real learning. But, *right now*, you have a specific objective to attain. This is not the time for daydreaming.

Step 4. Deal with your worries. Those pressing matters you can do something about right now, take care of. Those you can't do anything about, push aside for now.

Step 5. Don't *worry* about the exam. Take the exam seriously, but don't fret. Instead of worrying about the prospect of failure, use your time to eliminate failure as an option.

Plan

Let's go back to that general who knows the difference between objectives and goals. Chances are he or she also knows that you'd better not march off to battle without a plan. Remember, you don't want to *read.* You *want* to study. This requires focusing your work with a plan.

The first item to plan is your time.

Step 1. Dedicate a notebook or organizer-type date book to the purpose of recording your scheduling information.

Step 2. Record the following:

a. Class times
b. Assignment due dates
c. Exam dates
d. Extracurricular commitments
e. Study time

TIP: If you are in doubt about what tasks to assign the highest priority, it is generally best to allot the most time to the most complex and difficult tasks and to get them done first.

Step 4. Inventory your various tasks. What do you have to do today, this week, this month, this semester?

Step 5. Prioritize your tasks. Everybody seems to be grading you. Now's *your* chance to grade the things they give you to do. Label high-priority tasks "A," middle-priority tasks "B," and lower-priority tasks "C." This will not only help you decide which things to do first, it will also aid you in deciding how much time to allot to each task.

Step 6. Enter your tasks in your scheduling notebook and assign order and duration to each according to its priority.

Step 7. Check off items as you complete them.

Step 8. Keep your scheduling book up to date. Reschedule whatever you do not complete.

Step 9. Don't be passive. Actively *monitor* your progress toward your objectives.

Step 10. Don't be passive. Arrange and rearrange your schedule to get the most time when you need it most.

Packing Your Time

Once you have found as much time as you can, pack it as tightly as you can.

Step 1. Assemble your study materials. Be sure you have all necessary textbooks and notes on hand. If you need access to library reference materials, study in the library. If you need access to reference materials on the Internet, make sure you're at a computer.

Step 2. Eliminate or reduce distractions.

Step 3. Become an efficient reader and note taker.

An Efficient Reader

Step 3 requires further discussion. Let's begin with the way you read.

Nothing has greater impact on the effectiveness of your studying than the speed and comprehension with which you read. If this statement prompts you to throw up your hands and wail, "I'm just not a fast reader," don't despair. You can learn to read faster and more efficiently.

Consider taking a "speed reading" course. Take one that your university offers or endorses. Most of the techniques taught in the major reading programs actually do work. Alternatively, do it yourself.

Step 1. When you sit down to read, try consciously to force your eyes to move across the page faster than normal.

Step 2. Always keep your eyes moving. Don't linger on any word.

Step 3. Take in as many words at a time as possible. Most slow readers aren't slow-witted. They've just been taught to read word by word. Fast and efficient readers, in contrast, have learned to read by taking in groups of words. Practice taking in blocks of words.

Step 4. Build on your skills. Each day, push yourself a little harder. Move your eyes across the page faster. Take in more words with each glance.

Step 5. Resist the strong temptation to fall back into your old habits. Keep pushing.

TIP: Are you a—ugh!—*vocalizer*? A vocalizer is a reader who, during "silent" reading, either mouths the words or says them mentally. Vocalizing greatly slows reading, often reduces comprehension, and is just plain tiring. Work to overcome this habit—*except* when you are trying to memorize some specific piece of information. Many people do find it helpful to say over a sentence or two in order to memorize its content. Just bear in mind that this does not work for more than a sentence or paragraph of material.

When you review material, consider skimming rather than reading. Hit the high points, lingering at places that give you trouble and skipping over the stuff you already know cold.

An Interactive Reader

Early in this chapter we contrasted passive reading with active studying. A highly effective way to make the leap from passivity to activity is to become an *interactive* reader.

TIP: The physical act of underlining actually helps you memorize material more effectively—though no one is quite sure why. Furthermore, underlining makes review skimming more efficient and effective.

Step 1. Read with a pencil in your hand.

Step 2. Use your pencil to underline key concepts. Do this consistently. (That is, *always* read with a pencil in your hand.) Don't waste your time with a ruler; underscore freehand.

Step 3. Underline *only* the key concepts. If everything seems important to you, then up the ante and underline only the absolutely *most* important passages.

Step 4. If you prefer to highlight material with a transparent marker (a Hi-Liter, for example), fine. But you'll still need a pencil or pen nearby. Carry on a dialog with your books by writing condensed notes in the margin.

Step 5. Put difficult concepts into your own words—right in the margin of the book. This is a great aid to understanding and memorization.

Step 6. Link one concept to another. If you read something that makes you think of something else related to it, make a note. The connection is almost certainly a valuable one.

Step 7. Comment on what you read.

TIP: Some students are reluctant to write in their textbooks, because it reduces resale value. True enough. But is it worth an extra five dollars at the end of the semester if you don't get the most out of your multi-thousand-dollar and multi-hundred-hour investment in the course?

Taking Notes

The techniques of underlining, highlighting, paraphrasing, linking, and commenting on textbook material can also be applied to your classroom and lecture notes.

Of course, this assumes that you have taken notes. There are some students who claim that it is easier for them to listen to a lecture if they *avoid* taking notes. For a small minority, this may be true; but the vast majority of students find that note taking is essential when it comes time to study for midterms and finals. This does not mean that you should be a stenographer or court reporter, taking down each and every word. To the extent that it is possible for you to do so, absorb the lecture *in your mind*, then jot down major points, preferably in loose outline form.

Become sensitive to the major points of the lecture. Some lecturers will come right out and tell you, "The following three points are key." That's your cue to write them down. Other cues include:

- Repetition. If the lecturer repeats a point, write it down.
- Excitement. If the lecturer's voice picks up, if his or her face becomes suddenly animated, if, in other words, it is apparent that the material is at this point of particular interest to the speaker, your pencil should be in motion.
- Verbal cues. In addition to such verbal elbows in the rib as "this is important," most lecturers underscore transitions from one topic to another with phrases such as "Moving on to . . ." or "Next, we will . . ." or the like. This is your signal to write a new heading.
- Slowing down. If the lecturer gives deliberate verbal weight to a word, phrase, or passage, make a note of it.
- Visual aids. If the lecturer writes something on a blackboard or overhead projector or in a computer-generated presentation, make a note.

TIP: No one can tell you just how much to write, but bear this in mind: Most lecturers read from notes rather than fully composed scripts. Four double-spaced typewritten or word-processed pages (about a thousand words *in note form*) represent sufficient *note* material for an hour-long lecture. Ideally, you might aim at producing about 75 percent of this word count in the notes you take—perhaps 750 words during an hour-long lecture.

Filtering Notes

Some students take neat notes in outline form. Others take sprawling, scrawling notes that are almost impossible to read. Most students can profit from *filtering* the notes they take. This most emphatically does not mean rewriting or retyping your notes. This is a waste of time. Instead, underscore the most important points, filtering out the excess. If you have taken notes on a notebook or laptop computer, consider arranging the notes in clear outline form. If you have handwritten notes, however, it probably isn't worth the time it takes to create a neat outline. Spend that time merely underlining or highlighting the most important concepts.

Build Your Memory

Just as a variety of speed-reading courses are available, so a number of memory-improvement courses, audio tapes, and books are on the market. It might be worth your while to scope some of these out, especially if you are planning to go into a field that requires the memorization of a lot of facts. In the meantime, here are some suggestions for building your memory:

- Be aware that most so-called memory problems are really learning problems. You "forget" things you never *learned* in the first place. This is usually a result of passive reading or passive attendance at lectures—the familiar in-one-ear-and-out-the-other syndrome.
- Memorization is made easier by two seemingly opposite processes. First, we tend to remember information for which we have a context. Students in old-fashioned history classes quite rightly grouse about having to remember "a bunch of dates." Why is this so hard? Because the dates are too often taught (or learned) out of context, as a meaningless list rather than as part of the dramatic story of a series of interrelated historical events.
- Second, memorization is often made easier if we break down complex material into a set of key phrases or words. You may find it easier to memorize the key points relating to the Bill of Rights than to remember, say, the details of the story of how the Bill of Rights came into existence.
- It follows, then, that the best way to build your memory where a certain subject is concerned is to try to understand information in context. Get the "big picture."
- It also follows that, even if you have the big picture, you may want to break down key concepts into a few key words or phrases.

TIP: Make certain that the instructor approves of notebook computers and laptop computers as note-taking devices in class. Most lecturers have no problem with these, but some find the tap-tap-tapping of maybe more than a hundred students distracting.

Should you bring a tape recorder to class? The short answer is, probably not. To begin with, some instructors object to having their lectures recorded. Even more important, however, is the tendency to complacency a tape recorder creates. You might feel that you don't have to listen very carefully to the "live" lecture, since you're getting it all on tape. This is a mistake, since the live presentation is bound to make a greater impression on you, your mind, and your memory than a recorded replay.

How We Forget

It is always better to keep up with classwork and study than it is to fall behind and desperately struggle to catch up. This said, it is nevertheless true that most forgetting occurs within the first few days of exposure to new material. That is, if you "learn" 100 facts about Subject A on December 1, you may forget twenty of those facts by December 5 and another ten by December 10, but by March 1 you may still remember fifty facts. The curve of your forgetting tends to flatten.

Eventually, you may forget all 100 facts, but you forget them at the rate of twenty per week.

Now, what does this mean to you?

It means that you need to review material you learned earlier in the course. You cannot depend on having mastered it two, three, four, or more weeks earlier.

TIP: Many memory experts suggest that you try to put the key terms you identify into some sort of sentence; then memorize the sentence. Others suggest creating an acronym out of the initials of the key words or concepts. No one who lived through the Watergate scandal in the 1970s can forget that President Nixon's political campaign was run by CREEP (Committee to *RE*-Elect the President), and even nonscience majors remember how biologists traditionally classify living things by memorizing a sentence such as this: *Ken Put Candy On Fred's Green Sofa* (kingdom, phylum, class, order, family, species).

The Virtues of Cramming

Ask any college instructor about last-minute cramming for an exam, and you'll almost certainly get a knee-jerk condemnation of the practice. But maybe it's time to think beyond that knee jerk.

Let's get one thing absolutely straight. You cannot expect to pack a semester's worth of studying into a single all-nighter. It just isn't going to happen. However, cramming can be a valuable *supplement* to a semester of conscientious studying.

- You forget the most within the first few days of studying. (Or have you forgotten?) Well, if you cram the night before the exam, those "first few days" won't fall between your studying and the exam, will they?
- Cramming creates a sense of urgency. It brings you face to face and toe to toe with your objective. Urgency concentrates the mind.
- Assuming you aren't totally exhausted, material you study within a few hours of going to bed at night is more readily retained than material studied earlier in the day.

Burning the midnight oil may not be such a bad idea.

Cramming Cautions

Then again, staying up late before a big exam may not be such a hot idea, either. Don't do it if you have an early-morning exam. And don't transform cramming into an all-nighter. You almost certainly need some sleep to perform competently on tomorrow's exam.

Remember, too, that while cramming creates a sense of urgency, which may stimulate and energize your study efforts, it may also create a feeling of panic, and panic is never helpful.

Cramming is *not* a substitute for diligent study throughout the semester. But just because you *have* studied diligently, don't shun cramming as a *supplement* to regular study, a valuable means of refreshing the mind and memory.

TIP: Any full-time college student studies several subjects each semester. This makes you vulnerable to interference—the possibility that learning material from one subject will interfere with learning material in another subject. Interference is usually at its worst when you are studying two similar or related subjects. If possible, arrange your study time so that work on similar subjects is separated by work on an unrelated one.

Polly Want a Cracker?

We've been talking a lot about memory and memorization. It's an important subject and, for just about any course of study, an important skill. Some subjects—American history included—are more fact-and-memory intensive than others. However, beware of relying too much on simple, brute memory. Try to assess what the professor really wants: Students who demonstrate on exams that they have "absorbed" the facts the professor and the textbooks have dished out, or students who demonstrate such skills as critical thinking, synthesis, analysis, imagination, and so on. Depending on the professor's personal style and the kind of exam he or she gives (predominantly essay vs. predominantly multiple choice, for example), you may actually be penalized for "parroting" lectures. ("I know what *I* think. I want to know what *you* think.")

TIP: If you *hate* cramming, don't do it. It's not for you, and it will probably only raise your anxiety level. Get some sleep instead.

Use *This* Book (and Get Old Exams)

One way to judge what the professor values and expects is to pay careful attention in class. Is discussion invited? Or does the course go by the book and by the lecture? Also valuable are previous exams. Many professors keep these on file and allow students to browse through them freely. Fraternities, sororities, and formal as well as informal study groups sometimes maintain such files, too. These days, previous exams may even be posted on the university department's Wide World Web site.

Of course, you are holding in your hand a book chock full of sample and model midterm and final exams. Read them. Study them. And let them focus your study and review of the course.

Study Groups: The Pro and The Con

In the old days (whenever that was), it was believed that teachers *taught* and students *learned.* More recently, educators have begun to wonder whether it is possible to *teach* at all. A student, they say, *learns* by teaching him- or herself. The so-called teacher (who might better be called a "learning facilitator") helps the student teach him- or herself.

Well, maybe this is all a matter of semantics. Is there really a difference between *teaching* and *facilitating learning*? And between *learning* and *teaching oneself*? The more important point is that the focus in education has turned away from the teacher to the student, and students, in turn, have often responded by organizing study groups, in which they help each other study and learn.

These can be very useful:

- In the so-called real world (that is, the world after college), most problems are solved by teams rather than individuals.
- Many people come to an understanding of a subject through dialogue and question and answer.
- Studying in a group (or even with one partner) makes it possible to drill and quiz one another.
- In a group, complex subjects can be broken up and divided among members of the group. Each one becomes a "specialist" on some aspect of the subject, then shares his or her knowledge with the others.
- Studying in a group may improve concentration.
- Studying in a group may reduce anxiety.

Not that study groups are without their problems:

- All too often, study groups become social gatherings, full of distraction (however pleasant) rather than study. This is the greatest pitfall of a study group.
- All members of the group must be highly motivated to study. If not, the group will become a distraction rather than an aid to study, and it is also likely that friction will develop among the members, some of whom will feel burdened by "freeloaders."
- The members of the study group must not only be committed to study, but to one another. Study groups fall apart—bitterly—if members, out of a sense of competition, begin to withhold information from one another. This *must* be a Three Musketeers deal—all for one and one for all—or it is worse than useless.
- The group may promote excellence—or it may agree on mediocrity. If the latter occurs, the group will become destructive.

In summary, study groups tend to bring out the members' best as well as worst study habits. It takes individual and collective discipline to remain focused on the task at hand, to remain committed and helpful to one another, to insist that everyone shoulder their fair share, and to insist on excellence of achievement as the only acceptable standard—or, at least, the only valid reason for continuing the study group.

Thinking Like a Historian

As we just observed, instructors take varying approaches to the teaching of American history. Some emphasize the memorization of facts, while others stress original thinking, and while still others want something of both. All history professors, however, tend to look favorably on students who can think like historians. This includes:

- Demonstrating an eye for detail and respect for accuracy.
- Supporting statements of opinion with historical facts.
- Supporting judgments with appropriate quotations or statistics.
- Demonstrating sound inductive reasoning; that is, the ability to draw general conclusions from specific facts.
- Showing an ability to relate one fact or event to others in order to assemble the "big picture."
- Revealing an eye for cause and effect; much of a historian's work is determining what factors caused which effects.
- Developing an ability to evaluate divergent points of view.
- Possessing a good memory, which puts the most relevant facts at the student's fingertips.

CHAPTER 3

SECRETS OF SUCCESSFUL TEST TAKING

SOMETIMES IT SEEMS THAT THE DIFFERENCE BETWEEN ACADEMIC SUCCESS AND something less than success is not smarts versus non-smarts or even study versus nonstudy, but simply whether or not a person is "good at taking tests." That phrase—"good at taking tests"—was probably first heard back when the University of Bologna opened for business late in the eleventh century. The problem with phrases like this is that they are true enough, yet not very helpful.

Fact: Some people *are* and some people *are not* "good at taking tests."

So what? Even if successful test-taking doesn't come to you easily or naturally, you can improve your test-taking skills. Now, if you happen to have a knack for taking tests, well, congratulations! But that won't help you much if you neglect the kind of preparation discussed in the previous chapter.

Why Failure?

In analyzing performance on most tasks, it is generally better to begin by asking what you can do to succeed. But, in the case of taking tests, success is largely a question of *avoiding* failure. So let's begin there.

When the celebrated bank robber Willy "The Actor" Sutton was caught, a reporter asked him why he robbed banks. "Because that's where they keep the money," the handcuffed thief replied. At least one answer to the question of why some students perform poorly on exams is just as simple: "Because they don't know the answers."

There is no "magic bullet" in test taking. But the closest thing to one is *knowing the course material cold.* Pay attention, keep up with reading and other assignments, attend class, listen in class, take effective notes—in short, follow the recommendations of the previous chapter—and you will have taken the most important step toward exam success.

Yet have you ever gotten your graded exam back, with disappointing results, read it over, and found question after question you now realize you *could* have answered correctly?

"I *knew* that!" you exclaim, smacking yourself in the forehead.

What happened? You *really* did prepare. You *really* did know the material. What happened?

TIP: Most midterm and final exams really are representative of the course. If you have mastered the course material, you will almost certainly be prepared to perform well on the examination. Very few instructors purposely create deceptive exams or trick questions or even questions that require you to think beyond the course. Most instructors are interested in creating exams that help you and them evaluate your level of understanding of the course material. Think of the exams as natural, logical features of the course, not as sadistic assignments designed to trip you up. Remember, your success on an examination is also a measure of the instructor's success in presenting complex information. Very few teachers can—or want to—build careers on trying to *fail* their students.

Anxiety Good and Bad

The great American philosopher and psychologist William James (1842-1910) once advised his Harvard students that an "ounce of good nervous tone in an examination is worth many pounds of . . . study." By "good nervous tone" James meant something very like anxiety. You *should not* expect to feel relaxed just before or during an exam. Anxiety is natural.

Anxiety is "natural" because it is helpful. "Good nervous tone," alert senses, sharpened perception, adrenalin-fueled readiness for action are *natural* and *healthy* responses to demanding or threatening situations. We are animals, and these are reactions we share with other animals. The mongoose that relaxes when it confronts a cobra is a dead mongoose. The student who takes it easy during the American History final . . . Well, the point is neither to fight anxiety nor to fear it. Accept it, and even welcome it as an ally. Unlike our hominid ancestors of distant prehistory, we no longer need the biological equipment of anxiety to help us fight or fly from the snapping saber teeth of some animal of prey, but everyday we do face challenges to our success. Midterms and finals are just such challenges, and the anxiety they provoke is real, natural, and unavoidable. It may even help us excel.

What good can anxiety do?

- Anxiety can focus our concentration. It can keep the mind from wandering. This makes thought easier, faster, and, often, more acute and effective.
- Anxiety can energize us. We've all heard stories about a 105-pound mother who is able to lift the wreckage of an automobile to free her trapped child. This isn't fantasy. It really happens. And just as adrenalin can provide the strength we need when we need it most, it can enhance our ability to think under pressure.
- Anxiety moves us along. We work faster than when we are relaxed. This is valuable since, in most midterms and finals, limited time is part of the test.
- Anxiety prompts us to take risks. We've all been in classes in which the instructor has a terrible time trying to get students to speak and discuss and venture an opinion. "Come on, come on," the poor prof protests, "this is like pulling teeth!" Yet, when exam time comes, all heads are bent over bluebooks or notebook computers, and the answers—some kind of answers—are flowing forth or, at least, grinding out. Why? Because the anxiety of exam situation overpowers the inertia that keeps most of us silent most of the time. We take the risks we have to take. We answer the questions.
- Anxiety can make us more creative. This is related to risk taking. "Necessity," the old saying goes, "is the mother of invention." Phrased another way: *we do what we have to do.* Under pressure, many students find themselves taking fresh and creative approaches to problems.

So don't shun anxiety. But, unfortunately, the scoop on anxiety isn't all good news, either.

Anxiety evolved as a mechanism of *physical* survival. Biologists and psychologists talk of the "fight or flight" response. Anxiety prepares a threatened animal either to fight the threat or to flee from it. The action is physical and, typically, very short term. In our "civilized" age, the threats are generally less physical than intellectual and emotional, and they tend to be of longer duration than a physical fight or a physical flight. This means that the anxiety mechanism does not always work to enhance our chances for "survival" or, at least, our chances to survive the course by performing well on exams. Some of us are better than others at adapting the *physical* benefits of anxiety to the *intellectual* and *emotional* challenges of an exam. Some of us, unfortunately, are unable to benefit from anxiety, and, for still others of us, anxiety is downright harmful. Here are some of the negative effects anxiety may have on exam performance:

- Anxiety can make it difficult to concentrate. True, anxiety focuses concentration. But if it focuses concentration on the anxious feelings themselves, you will have less focus left over for the exam. Similarly, anxiety may cause you to focus unduly on the perceived consequences of failure.
- Anxiety causes carelessness. If anxiety can prompt you to take creative risks, it can also cause you to rush through material and, therefore, to make careless mistakes or simply to fail to think through a problem or question.
- Anxiety distorts focus. Anxiety may impede your judgment, causing you to give disproportionate weight to relatively unimportant matters. For example, you may become fixated on solving a lesser problem at the expense of a more important one. This is related to the next point.

TIP: Exam questions are battles in a war. No general expects to win every battle. Accept your losses and move on. Dwell on your losses, and you will continue to lose.

- Anxiety may distort your perception of time. You may think you have more or less of it than you really do. The result may be too much time spent on a minor question at the expense of a major one.
- Anxiety tends to be cumulative. Many test takers have trouble with a question early in the exam, then devote the rest of the exam to worrying about it instead of concentrating on the *rest of the exam.*
- Anxiety drains energy. For a short period of time, anxiety can be energizing and invigorating. But if anxiety becomes chronic, it begins to tire you out. You do not perform as well.
- Anxiety can keep you from getting the rest you need. If it is generally unwise to stay up all night *studying* for an exam, how much less wise is it to stay up uselessly *worrying* about one?

How can you combat anxiety?

Step 1: *Don't* fight it. Accept it. Remember, anxiety is a *natural* response to a stressful situation. Remember, too, that some degree of anxiety aids performance. Try to learn to accept anxiety and *use* it. Let it sharpen your wits and stoke the fires of your creativity.

Step 2: Don't worry about how you feel. Focus on the task. Usually, you will feel better once you overcome the initial jitters and inertia. William James, who lauded "good nervous tone," also once observed that we do not run because we are frightened, but we are frightened because we run. If you concentrate on your fear and act as if you are afraid, you will become even more fearful.

Step 3: Prepare for the exam. Do whatever you must do to master the material. Build confidence in your understanding of the course, and your anxiety should be reduced.

TIP: More serious stimulant drugs ("speed," "uppers") are *never* a good idea. They are both illegal and dangerous (possibly even deadly), and, for that matter, their effect on exam performance is unpredictable. The chances are they will impede performance rather than aid it (though you may erroneously *feel* that you are doing well). Also avoid over-the-counter stimulants. These are caffeine pills and will probably increase anxiety rather than improve performance.

Step 4: Get a good night's sleep before the exam.

Step 5: Avoid coffee and other stimulants. Caffeine tends to increase anxiety. (However, if you are a caffeine fiend, don't pick the day or two before a big exam to kick the habit. You *will* suffer withdrawal symptoms.)

Step 6: Try to get fresh air shortly before the exam. This is especially valuable if you have been cooped up for a long period of study. Take a walk. Get a look at the wider world for a few minutes.

Have a Plan, Make a Plan

A large component of destructive anxiety—probably the largest—is fear of the unknown. Reduce anxiety by taking steps to reduce the component of the unknown.

Step 1: To repeat—Do whatever is necessary to master the material on which you will be examined.

Step 2: Use the exams in this book to familiarize yourself with the kinds of exams you are likely to encounter.

Step 3: If possible, examine old exams actually given in the course.

Let's pause here before going onto Step 4. Just reading over the exams in this book or leafing through exams formerly given in the course will not help you much. Analyze:

- The types of questions asked. Are they essay or objective questions? (We'll discuss these shortly.) Do they call for "regurgitation" of memorized material, or are they more "think"-oriented, requiring significant initiative to answer?

- Don't just predict which questions you could and could not answer. Try actually answering some of the questions.
- If you are looking at sample exams with answers, evaluate the answers. How would you grade them? What would you do better?
- Don't just stand there, do something. If your analysis of the sample exams or old exams reveals areas in which you are weak, address those weaknesses.

An effective way to reduce the unknown is to create a plan for confronting it. Let's go on to Step 4.

Step 4: Make a plan. Begin *before* the exam. Decide what areas you need to study hardest. Based on your textbook notes and—especially—on your lecture notes, try to anticipate what kinds of questions will be asked. Work up answers or sketches of answers for these.

TIP: Don't make the mistake of devoting all of your time to trying to make last-minute repairs to weak spots ("I've got one hour to read that textbook I should have been reading all along!") only to ignore your strengths. Develop your strengths. With any luck at all, the exam will give you an opportunity to show yourself at your best—not just trip you up at your worst. Be as prepared as you can be, but, remember, there is nothing wrong with excelling in a particular area. Play to your strengths, not your weaknesses.

Step 5: Make sure you've done the simple things. The night before your exam, make certain that you have whatever equipment you'll need. If you will be allowed to use reference materials, bring them. If you are permitted to write on a laptop or notebook computer, make certain your batteries are fully charged. If you are writing the exam longhand, make certain you have pens, pencils, paper. Bring a watch.

Step 6: Expect a shock. The first sight of the exam usually packs a jolt. At first sight, questions may draw a blank from you. Questions you were sure would appear on the exam will be absent, and some you never expected will be staring you in the face. Don't panic. *Everybody feels this way.*

Step 7: Write nothing yet. Read through the exam thoroughly. Be certain that you: (a) understand any instructions and (b) understand the questions.

Step 8: If you are given a choice of which questions to answer, choose them now. Unless the questions vary in point value assigned to them, choose those that you feel most confident about answering. Don't challenge yourself.

Alternative Step 8: If you are required to answer all the questions, identify those about which you feel most confident. Answer these first.

TIP: When you study for an exam, it is usually best to assign high priority to the most complex and difficult issues, devote ample time to these, and master them first. When you *take* an exam, however, and you are under time pressure, tackle first what you can most readily and thoroughly answer, then go on to more doubtful tasks. Your professor will be more favorably impressed by good or correct answers than by failed attempts to answer questions you find difficult.

Step 9: After you have surveyed the exam, create a "time budget": note—jot down—how much time you should give to each question.

Step 10: Reread the question before you begin to write. Then plan your answer.

Plan Your Answer

Perhaps you have heard a teacher or professor comment on the exam he or she has just handed out: "The answers are in the questions." This kind of remark is more helpful than it may at first seem.

Begin by looking for the key words in the question. These are the verbs that *tell* you what to do, and they typically include:

- Compare
- Contrast
- Criticize
- Define
- Describe
- Discuss
- Evaluate
- Explain
- Illustrate
- Interpret
- Justify
- Outline
- Relate
- Review
- State
- Summarize
- Trace

You will find most of these key words in essay rather than objective or short-answer questions, so we will have much more to say about the key words in the next chapter, which is devoted to answering the essay exam. But be aware that the following key words are often found in short-answer questions:

- To ***define*** something is to state the precise meaning of the word, phrase, or concept. Be succinct and clear.

- To ***illustrate*** is to provide a specific, concrete example.
- To ***outline*** is to provide the main features or general principles of a subject. This need not be in paragraph or essay form. Often, outline format is expected.
- To ***state*** is similar to define, though a statement may be even briefer and usually involves delivering up something that has been committed to memory.
- To ***summarize*** is briefly to state—in sentence form— the major points of an argument or position or concept, the principal features of an event, or the main events of a period.

As has just been mentioned, advice on answering essay examinations is the subject of the next chapter. For the moment, just be aware that you will want to budget time for creating a scratch outline of your essay answer.

Approaching the Short-Answer Test

While the value of planning is more or less obvious in the case of essay exams, you will also find it valuable in "objective," multiple-choice, or other short-answer exams. These are of two major kinds:

1. *Recall exams* include questions that call for a single short answer (usually there is a single "correct" answer) and fill-in-the-blank questions, in which you are asked to supply missing information in a statement or sentence.
2. *Recognition* exams include multiple-choice tests, true-false tests, and matching tests (match one from column A with one from column B).

If the exam is a long one and time is short, invest a few minutes in surveying the questions, so that you can be certain to answer those you are confident of, even if they come near the end of the exam.

Be prepared to answer multiple-choice questions through a process of elimination, if necessary. Usually, even if you are uncertain of the one correct answer among a choice of five, you *will* be able to eliminate one, two, or three answers you know are *incorrect.* This at least increases your odds of giving a correct response.

Unless your instructor has specifically informed you that he or she is penalizing guesses (actually taking points away for incorrect responses versus awarding zero points to unanswered questions), do guess the answers even to those questions that leave you in the dark.

Plan your responses to true-false questions carefully. Look for telltale qualifying words, such as *all, always, never, no, none,* or *in all cases.* Questions with such absolute qualifiers often require an answer of *false,* since relatively few general statements are always either true or false. Conversely, questions containing such qualifiers as *sometimes, usually, often* and the like are frequently answered correctly with a response of *true.*

A final word on guessing: *First guess, best guess.* Statistical evidence consistently shows that a first guess is more likely to be right than a later one. Obviously, if you have responded one way to question and then the correct answer suddenly dawns on you, do change your response. But if you can choose only from a variety of guesses, go with your first or "gut" response.

Take Your Time

Yes, yes, yes, this is easier said than done. But the point is this:

- Plan your time.
- Work efficiently, not in a panic.
- Make certain that your responses are legible.
- Take time to spell correctly. Even if an instructor does not consciously deduct points for misspelling, such basic errors will negatively influence the evaluation of the exam.
- If a short answer is called for, make it short. Don't ramble.
- Use *all* of the time allowed. The instructor will *not* be impressed by a demonstration that you have finished early. If you have extra time, reread the exam. Look for careless errors. Do not, however, heap new guesses on top of old ones; where you have guessed, stick with your first guess.

Essay Exams: Read On

Please continue to the next chapter for advice on preparing for and writing effective essay responses.

CHAPTER 4

THE ESSAY EXAM: WRITING MORE EFFECTIVE RESPONSES

WHILE MANY EXAMS IN INTRODUCTORY- AND SURVEY-LEVEL HISTORY COURSES include multiple-choice and fill-in-the-blanks segments, most emphasize essay questions. Many include short-answer essays —questions requiring a paragraph or two in response—as well as longer essays, which may consume several pages of an exam "blue book."

Downside and Up

Griping is of little value at the outset of any enterprise, including a history course replete with essay exams. But griping is human and natural, so we'd better get it out of the way. Essay exams have the following distinct disadvantages:

1. They are intimidating. Even experienced, professional writers may get a shudder when they sit down to a blank page. Where do you begin? Worse, where do you go once you've begun?
2. Essay questions generally require deeper and broader knowledge of a subject than multiple-choice questions do.
3. Essay exams are time-consuming to take. It may be difficult to budget your time effectively.
4. Essay exams test not only your knowledge of course material, but your language and writing skills. This may seem unfair.

5. Essay questions contain a significant element of subjectivity. Typically, the issues are gray rather than black and white. Not only does subjectivity enter into your response, it also plays a role in the instructor's evaluation of the response. An instructor may respond to the skill of the presentation (or lack of such skill) as much as she does to the substance of the answer.

In fairness, essay exams are almost as demanding on the instructor as they are on the students. They are much more difficult and time-consuming to grade than "objective" tests are. The instructor who uses essay exams is demonstrating a genuine commitment to her students and her subject.

So there's the downside. But each of these negatives has a corresponding positive—if you know how to find and exploit it. Look:

1. True, your blue book may be blank, but your mind doesn't have to be. First, there are effective ways to prepare for the questions on an essay exam—and we'll talk about these shortly. Second, take a good, long, careful look at the question. It should give you plenty to get you started. Get into the habit of using the terms, parts, and structure of the question as a kind of framework on which you construct the terms, parts, and structure of your answer.
2. If essay questions usually require deeper and broader knowledge of a subject than multiple-choice questions do, they also offer a deeper and broader stage on which you can play out your understanding of the course material. When you respond to short-answer and multiple-choice questions, you are limited by the instructor's rules: true, false, a, b, c, or d. It's a kind of binary situation. Either you *know* the *correct* answer or you don't—and if you don't, you lose. In responding to an essay question, you certainly need to address the question in all of its parts (don't stray, don't evade, don't get off the track), but you have much more control. You can focus on areas you know most about. You can play to your strengths and minimize your weaknesses.
3. Essay exams are time-consuming to take. That's a fact. And your instructor knows it. He takes into account the pressure of limited time and the fact that you are writing a single draft when he evaluates the essay. This generally prompts him to overlook a lot of sins of omission and even outright errors. Time pressure actually reduces your instructor's expectations.
4. If essay exams test not only your knowledge of course material, but also

your language and writing skills, it behooves you to polish those skills. Good writing will earn extra points. It's that simple. You may not know more about history than the person sitting next to you, but if you write more effectively, you will earn a higher grade. An added bonus: The more clearly and effectively you can express yourself, the *better* your own understanding of the material you are writing about will be. Effective writing not only communicates knowledge to others, it helps you to communicate with yourself.

5. Essay questions contain a significant element of subjectivity, it is true, and this can give you that much more "room" to be right. Create a skillful presentation, and you are likely to be evaluated positively, even if you miss some issues.

Study and Preparation

Even if they deny it, most instructors tend to "teach for the midterm and final." More accurately put, they construct exams that genuinely reflect the course content, including particular themes and topics that are emphasized. Do instructors ask "trick questions"? Rarely. Do they deliberately try to mislead you, hiding exam material in the background of the course, as if it were an Easter egg? Almost never. The fact is that instructors want you to succeed. Good test performance tells them that they have gotten through to you. The more students who do well, the more successful an instructor feels. This being the case, be certain to take careful lecture notes. Make a good set of general notes, but also listen selectively for:

TIP: Some instructors prepare examinations well ahead of time, but most write them up shortly before they are given. Usually, the instructor will review her notes in preparation for writing an exam. This makes it even more likely that well-emphasized subjects and issues will appear on the exam.

- Points that stand out
- Points that are repeated
- Points preceded by such statements as, "Now this is important" or "This is a major issue" and the like.

Assume that any point, theme, or topic that is given special emphasis will appear as an exam topic. The more emphasis it is given, the more likely it is to appear as an essay exam question.

Typically, course lectures mesh with textbook assignments, additional assigned outside reading, and perhaps class handouts. Take notes on all of these sources. Handouts are usually especially important. Assume that handout

TIP: Do not confine your preparation to your notes. If past course exams are available for your review, review them. Use them as practice tests. Of course, you should also make use of the exams in this book. These days, many instructors post past exams on a special web site devoted to the course.

material will figure in some way on any exam.

Avoid passivity. *Ask* the instructor to talk about the scope of the exam. Also seek out students who have taken the course before. Ask them about the exam. Many instructors have favorite themes or concerns, which get repeated from year to year.

You Are Not Alone

You needn't face the exam alone. Group study is often a highly effective way of preparing for essay exams. Indeed, many instructors encourage students to form study groups. This is because group discussion tends to bring out major themes and issues—the meat and potatoes of essay exams—rather than mere facts, which are the focus of short-answer or multiple-choice exams.

Don't make the mistake of allowing the study group to dissolve into a social hour. Consider focusing the discussion by having each member of the group make up a sample essay question. Use these as the topic of discussion.

Limited Possibilities

Let's assume you're not musically inclined. Now, look at the score of, say, the Beethoven *Moonlight Sonata*. All those notes! All those chords! How does anyone ever learn how to read so much simultaneous information, much less translate it into sound on a keyboard?

Well, it requires study, hard work, practice—and inborn musical talent helps, too. But it really isn't as hard as it appears to the unmusical. While, theoretically, there are an infinite number of ways in which musical notes can be combined and deployed, in actual practice, the possibilities are indeed limited. Most chords and note sequences occur in recognizable groups and patterns. Just as, when you read a book, you don't spell out individual words or struggle to recognize individual letters, but instead more or less unconsciously interpret familiar linguistic patterns and phrases, so a musician processes the notes he sees.

PITFALL: By all means, examine tests from previous semesters, but don't make the mistake of assuming that these will simply be repeated this semester. Most instructors change tests from year to year. Examine past tests to get an idea of the type and scope of questions asked—not to get specific answers.

Now, you may think that the range of questions possible in an essay exam is virtually infinite, but like the notes of a musical score or the words on a novel's page, the range is limited—and this is true regardless of subject.

There are a limited number of questions that can be asked about any theme, subject, or topic. Furthermore, each question is controlled by a key word.

Knowing those key words—and understanding their meaning—will help you to prepare adequately for the exam. Here they are:

Analyze Literally, take apart. Break down a subject into its component parts and discuss how they relate to one another.

Compare Identify similarities and differences between (or among) two (or more) things. End by drawing some conclusion from these similarities and differences.

Contrast Set two (or more) things in opposition in order to bring out the differences between (or among) them. Again, draw some conclusion from these differences.

Criticize Make a judgment on the merits of a position, theory, opinion, or interpretation concerning some subject. Support your judgment with a discussion of relevant evidence.

Defend Give one side of an argument and offer reasons for your opinion.

Define State as precisely as possible the meaning of some word, phrase, or concept. Develop the definition in detail.

Describe Give a detailed account of something. Where historical subjects are concerned, this account will typically be chronological.

Evaluate Appraise something, rendering a judgment as to its truth, usefulness, worth, or validity. Support your evaluation with relevant factual evidence.

Explain Clarify something and provide reasons for it.

Identify Define or characterize names, terms, places, or events.

Illustrate Provide an effective example of some stated point, principle, or concept.

Outline Show the main features of some event, concept, idea, political movement, and so on. Omit the details. Often, such an answer will be in outline rather than narrative form.

Pros and cons A more specialized form of evaluate. List and discuss the positives and negatives about a certain position, idea, event, movement, and so on.

Relate Narrate an event or set of events; emphasize the relation of one event to another, especially focusing on cause and effect.

Review Survey a subject. This is much like an *outline*, but put in narrative rather than list or outline form.

State Present your answer briefly and clearly, usually using a simple declarative sentence: *Such-and-such is such-and-such.*

Summarize Give a concise account of major points, ideas, or events. Skip details and examples.

Trace Follow an event, movement, or idea to its origin. The form of this response is generally: "The origin of A is X, Y, Z." Then the rest of the answer continues by elaborating on X, Y, and Z and moves forward to A.

One or more of these are the intellectual operations basic to just about any essay. Look for these key words in the essay question.

Many of these operations focus on just two elements:

- cause
- effect

Indeed, most history essay questions deal with causes and effects. Be prepared in advance to work with both elements.

TIP: You can use the operations and elements just discussed to focus your study notes. For example, instead of listing a bunch of unrelated facts about the Great Depression, why not focus your notes in terms of causes and effects and also *trace* the Depression to its root causes. You might then study FDR's New Deal by *evaluating* it or listing its *pros and cons.*

Test Time: The Problem of Inertia

If you've just finished up an exam in Physics 101, maybe you recall what Sir Isaac Newton had to say about inertia—the tendency of a body in motion to remain in motion and a body at rest to remain at rest. Sitting down to an essay question on exam day, many students are confronted by their own personal form of inertia. You look at the question. And look at it. And look at it some more, hoping, perhaps, that the letters will magically rearrange themselves on the page to yield up the answer.

The bad news, of course, is that they will do no such thing. But the good news is that the germ of the answer is, in fact, in the question. How do you overcome essay test inertia—that paralyzing difficulty in getting started? Just read the question. *Really* read the question.

In fact, don't worry about writing just now. Sit down and read *all* of the questions before you begin to write anything.

Here's why:

- If you are given a choice of which questions to answer (say two out of three), you want to be sure that you answer the ones you know best.
- Questions are sometimes related. You want to get an idea of just how they are related to one another, so that you don't "waste" too much of your answer on one question to the neglect of another.
- You need to assess your time needs. Are there some points that will require more time than others?
- You want to make certain that you answer the questions you are confident about first. Given a limited amount of time, be certain that you get to your best shots first and complete them before attending to the questions you're less confident about.

TIP: People who give a lot of speeches are fond of offering this formula for a successful speech—"Tell them what you are going to say. Say it. Tell them what you said." You might keep this in mind when you are answering an essay question, though you should elaborate on the formula a bit:

1. **State your subject or thesis.**
2. **Briefly state how you will discuss it. (A, B, C, D—or maybe C, A, B, D).**
3. **Answer the question.**
4. **Concisely summarize your answer.**
5. **Draw any additional conclusions as to significance, ramifications, etc.**

Before you begin to write, be absolutely certain that you understand each question completely. Read each question actively, aggressively, with pencil in hand:

- Identify and underline key words—including those listed above as basic "operations" and "elements."
- Be certain to *do* what the key words ask. For example, don't just *define* when you are asked to *explain.*
- If a question is complex, consisting of several subquestions, make certain you understand and answer all parts of the question.

Generally, you should let the question provide the basic structure for your response. If a question consists of subquestions A, B, C, and D, begin by answering A, then B, C, and D. If you have a *very good* reason for changing the order of your response, be certain to explain it. For example, you might begin: "Because C is essential to understanding A, B, and D, I will begin by discussing C, then proceed to A, B, and D."

Finally, answer the question—and only the question. Make certain that you address all parts of the question, but don't go beyond what the question asks—

unless, after you have thoroughly addressed each aspect of the question, you feel it is important to bring in additional issues. If you do so, tell your reader what you are doing, so that he won't think you've misunderstood the question and are simply going off on a tangent.

To Outline or Not to Outline

Sometimes, pulling out ideas in response to an essay question is difficult, halting, and laborious. Sometimes, however, you are flooded with ideas. Either situation can make inertia more powerful and, at worst, bring on mental paralysis or, at the very least, cause you to write a poorly organized essay. To prevent these outcomes, apportion your time so that you spend about half of the exam period planning and outlining your response.

Now, an essay exam outline does not have to be a formal outline with major and minor headings. Perhaps a simple list will be sufficient. Just make a map of your answer, setting down the main points that need to be covered. This will have three effects:

1. It will help ensure that you leave out nothing important.
2. It will help you organize the logic of your response.
3. It will reduce your anxiety.

The first two points are obvious. The last is less obvious, but no less important. Without an outline, you may fear that you will forget something important or get lost in your response. Get the main points out of your head and onto paper quickly, and you won't have to worry about forgetting anything or getting lost.

Structure Strategy

Make the structure of your essay exam response as clear and obvious as possible. Begin with a thesis statement: a statement or listing of the main idea you will support and develop in the body of the essay.

Where do you get your thesis statement?

The first place to look is the question. "President James Buchanan has been criticized for doing nothing to prevent the Civil War. Do you agree or disagree? Explain and defend your answer." There's the question. Begin by getting to the point: "President Buchanan failed to take action to avert the Civil War." That is an effective thesis statement. It should lead you to explain how you intend to support the statement: "I will explain just what he failed to do and suggest what he might have done."

In general, make the thesis statement as simple and as direct as possible. State the thesis. Present your plan for supporting your thesis. Follow your plan in the body of the essay:

I. Buchanan did nothing to prevent the Civil War

II. What he failed to do

A. Failure A
B. Failure B
C. Failure C

III. What he could have done
A. Step A
B. Step B
C. Step C

IV. Conclusion: Buchanan's critics are justified

The K.I.S.S. Formula

Can you really get ahead by giving your professor a KISS? Well, sort of. This acronym stands for *K*eep *I*t *S*imple, *S*tupid. Now, let's get something straight. This does not mean that you should overly simplify complex ideas or issues, let alone avoid them. But it does mean that you should structure your presentation of ideas in as simple a form as *possible.* This means:

- Be concise.
- Be direct.
- Start the essay with a thesis statement.
- Start each paragraph with a topic sentence, announcing the subject of the paragraph.
- Try to make a single major point in each paragraph.
- When you move on to a new point or new idea, start a new paragraph. Essay exam essays usually have short paragraphs—certainly shorter than what you'd write in a term paper, for example.
- Draw a definite conclusion.

Specify, Always Specify

Wherever possible, avoid abstraction. In place of vague generalizations, make very specific points that use specific examples. Examples are important in any essay response, and they are especially important in history essays.

Sign Posts

Develop a repertoire of verbal sign posts. One of the most effective sign posts is enumeration. For example, instead of saying "Several factors contributed to the U.S.-Mexican War," write "Five major factors contributed to the U.S.-Mexican War." Then go on to list and discuss all five factors.

Enumeration accomplishes three things:

1. It creates the impression that you are in control of the information.
2. It creates an impression of precision and completeness.
3. It sets up an expectation in the reader, who is satisfied when that expectation is fulfilled. You promise five items. You deliver five items. The reader is impressed.

Other sign posts include words and phrases such as:

To begin with . . .
First . . .
Next . . .
Therefore . . .
If . . . then
Because . . .
The result of . . .
. . . caused by . . .
However . . .
Except for . . .
Including . . .
For example . . .
Although . . .
Since . . .
Consequently . . .
Finally . . .
In conclusion . . .

Use these to get your reader from one point to the next, to make clear exceptions, and to point your reader toward your conclusions.

A Few Words on Words

Take time to choose your words carefully. This does not mean trying to impress the instructor with big words or fancy words, but do try to find the right words, the words that most precisely say what you mean.

- Use the language of history. Identify the specialized political, geographical, and economic terms used in lectures and textbooks. Understand these terms. Use them appropriately in writing the exam essays.
- Avoid slang. Slang is not only imprecise, it creates a poor impression.
- Prefer strong nouns and verbs to adjectives and adverbs. This will help you to be more precise: "The Viet Nam War was terribly bloody" is a far weaker statement than "The Viet Nam War cost more than 50,000 American lives."
- Express yourself in a direct manner. Don't load up sentences with unnecessary words. Try to make each word count. Use the active voice instead of the passive voice: "John Wilkes Booth assassinated President Lincoln" is a much stronger sentence than "President Lincoln was assassinated by John Wilkes Booth."

TIP: Try to budget time to reread and proofread your essay. Catch and correct errors of usage, grammar, and spelling.

- Avoid unnecessary qualifying phrases and waffling words such as "it has been said" or "I think" or "it seems to me" or "it seems likely that." Make direct statements.
- Avoid padding and repetition: "President Johnson was troubled by many troubling problems, which burdened him greatly."
- Avoid errors of usage, grammar, and spelling. If you have trouble in these areas, work on them. Such errors undercut your credibility.

Neatness Counts

Take enough time to write legibly. Make your work as easy to read as possible. If the instructor has to struggle to decipher your handwriting, he will easily lose the thread of your discussion, and your grade is likely to suffer as a result.

Recycle!

When the graded exam comes back to you, resist the temptation either to pat yourself on the back or to kick yourself in the backside. Instead, carefully read the examiner's comments. Learn from them. Schedule a conference to discuss the exam—*not* with the goal of getting your grade changed, but of identifying those areas that can use improvement.

When you have a conference, try not to respond defensively. Invite frank feedback. Don't get offended or upset. Instead, look for patterns. Does the instructor say that you simply failed to answer the question adequately? That you didn't answer all parts of the question? That you answered vaguely? That you punctuated poorly? That you didn't use enough examples? Diagnose areas that need improvement, even as you identify your strengths. Your ultimate goal is to avoid repeating errors while working to duplicate your successes.

PART TWO

STUDY GUIDE

CHAPTER 5

U.S. HISTORY: THE MAJOR THEMES

WHILE INTRODUCTORY OR SURVEY-LEVEL AMERICAN HISTORY COURSES IN A FEW colleges and universities cover the entire span of American history—that is, from early exploration and settlement through the present—most are divided into two courses, one typically beginning with the earliest period and going through the Civil War (that is, 1861-65) or Reconstruction (ending in 1877), and the other taking up where the first left off, then continuing through the present.

Some professors structure their courses around leading themes, such as the development of democracy, the pursuit of happiness, slavery and its consequences, the immigrant experience, or the evolution of the United States as a world power. Most courses, however, are structured chronologically. While various intellectual or social themes emerge, develop, and are repeated, most history courses consist of chronological "themes."

The following are some of the broad subject areas you can expect to be addressed in the first of a typical two-course American history survey.

Colonial America Before 1763

- The "Native" Americans
- European Exploration
- Spanish Settlement

- English Settlement
- Other Immigrant Groups
- Colonial Cultures
- Colonial Economics
- Special Significance of Puritan New England

Era of Revolution (1763-1789)

- The French and Indian Wars
- The French and Indian War
- Roots of Revolution
- The Taxation Crisis
- The Revolution Begins
- Revolution on the Seaboard
- European Allies
- Revolution on the Frontier

The Early Republic (1789-1824)

- Articles of Confederation
- Constitutional Convention
- Bill of Rights and Ratification
- Government under Washington and Hamilton
- Birth of the Party System
- Government Under Adams and Marshall
- The Age of Jefferson
- Louisiana Purchase
- War of 1812
- The "Era of Good Feelings"

The Age of Jackson and Westward Expansion (1824-1850)

- Rise of "the Common Man"
- Jackson vs. the Bank of the United States
- Nullification Crisis
- Indian Removal
- Age of Reform
- Slavery and Abolitionism
- Texas Independence
- Manifest Destiny and a War with Mexico

- Immigrant Trails and a Rush for Gold

The Era of Civil War and Reconstruction (1850-1877)

- Compromise of 1850
- "Bleeding Kansas"
- Birth of the Republican Party
- The Dred Scott Decision
- Emergence and Election of Lincoln
- Secession Crisis
- Civil War: Fort Sumter to Antietam
- Civil War: The Emancipation Proclamation
- Civil War: Gettysburg to Appomattox
- Assassination of Lincoln
- Reconstruction: Civil Rights and Civil Wrongs

At this point, with the conclusion of the Civil War and its bitter aftermath in Reconstruction, many "part one" introductory U.S. history courses end. A new America emerged beginning in the last quarter of the nineteenth century, with the influx of huge numbers of immigrants and the rise of urban industrialism, which steadily supplanted the rural, agrarian way of life in the United States. Major "part two" course themes include:

Industrialization, Populism, and Progressivism (1877-1916)

- The Era of the Immigrant
- Urbanization and Industrialization
- Robber Barons and Unionism
- The Rise of American Popular Culture
- More Westward Expansion
- The Indian Wars
- Populism
- Progressivism
- American Imperialism: The Spanish-American War
- The Era of Teddy Roosevelt
- Roosevelt's Imperialism
- The Conservation Movement
- Social Retreat Under Taft
- "New Freedom" Under Wilson

From World War I to World War II (1916-1945)

- Isolationism and its End
- America Enters the "Great War"
- Fourteen Points and a League of Nations
- The Lost Generation and the Roaring Twenties
- Prohibition and the Emergence of Organized Crime
- New Prosperity
- The Rise of Women
- The Crash of 1929 and the Great Depression
- FDR and the New Deal
- The U.S Enters World War II
- World War II: Pacific Theater
- World War II: European Theater
- The Atomic Age Is Born
- The Home Front

Postwar America (1945-1968)

- Social Revolutions
- Truman, Korea, and the Cold War
- The McCarthy Era
- The Civil Rights Movement
- The Eisenhower Years
- JFK's "New Frontier"
- Cuban Missile Crisis
- Assassination
- LBJ and the Great Society
- The Vietnam War

Recent History (since 1968)

- Space Race Culmination: Moon Landing
- Culture, Counterculture, and Drug Culture
- Nixon and China
- The Watergate Crisis
- OPEC and the Energy Crisis
- Reaganomics
- The Cold War Ends
- Iran-Contra Scandal

- Desert Storm
- New Directions

The work of the historian becomes more difficult as it deals with recent or current events. Contemporary "history" is, by definition, a work in progress, and it is therefore hard to draw conclusions as to the significance of various still-unfolding events. Often, history teachers approach current events in relation to the past, asking students to relate some current situation to one or more analogous situations in the past.

Your history course may or may not cover all of the themes presented here. Almost certainly, it will also include others we have *not* listed. But you can be confident that most of the themes listed here will figure prominently in any history survey.

CHAPTER 6

COLONIAL AMERICA BEFORE 1763

AS YOU WILL DISCOVER WHEN YOU REVIEW THE COMMENTS VARIOUS INSTRUCTORS make about the exams in this book, rote memorization of facts, events, people, and dates is far less important to success in a history course than is an understanding of major historical themes and how these relate to one another. If there is one overarching idea that binds the events of earliest American history, it is the New World—the Western Hemisphere—as a great stage on which three very different cultures met, interacted, and often conflicted. Europeans (chiefly of Spanish and British origin), Native Americans, and African slaves all figures prominently in early American history.

The "Native" Americans

We call the American Indians "native" Americans because they were here when the European explorers and settlers arrived. Yet, in the very distant past, anywhere from 10,000 to 50,000 years ago, the ancestors of the so-called Native Americans were themselves immigrants. It is believed that a "land bridge" once existed where the Bering Strait now is, and that Asian peoples gradually migrated across it, eventually spreading throughout North, Central, and South America. This process consumed thousands of years and resulted in far-flung populations of great cultural diversity who spoke over a thousand different languages. The life-styles of the Indians was also highly diverse, ranging from the elaborate and elaborately organized civilizations of Central America (the Maya and Aztec) and South America (the Inca) to the smaller, simpler societies of most North American groups.

Coverage of Native Americans varies greatly in introductory history courses. This is due in part to a lack of written records from the era before European contact. You may study various aspects of Native American *culture* rather than *history*—for example, agricultural practices, religion, division of labor between men and women, hunting patterns, and so on. One thing you will likely discover is that Native Americans did not necessarily lead peaceful lives before the coming of the Europeans. Intertribal warfare was quite common and, for many groups, a way of life.

European Exploration

The Crusades of the eleventh through fifteenth centuries introduced Europeans to the worlds of the Middle East and Asia, initiating trade relationships with peoples of these regions. During the Renaissance (fourteenth through sixteenth centuries), a new merchant class arose, as did such sources of investment capital as joint stock companies. During this period, too, the Protestant Reformation challenged Roman Catholicism, splintering Christianity and pitting various Christian sects in competition against one another. Finally, the Renaissance saw the emergence of powerful nations that possessed the resources and motivation to explore and to conquer.

Thus economic, political, religious, and cultural changes motivated Europeans to seek new worlds. In general, they hoped to find:

- Greater wealth—in trade, in gold, in spices (vital as food preservatives in an era before refrigeration), and in slaves
- National prestige and power—world dominance
- Religious fulfilment—both Protestant and Catholic nations were driven by strong missionary impulses; they wished to spread their religion to new regions

Moreover, opportunities were limited in Europe. Among the nobility, the practice of primogeniture (in which the first son in a family inherits all family titles and all or most of the family's wealth) left many young men with few prospects at home. As for the emerging merchant classes, they were always eager for new markets for their wares. Lower-class people were looking for alternatives to dire poverty.

Although it was probably Scandinavian (Norse or Viking) explorers who first ventured to the New World (as early as A.D. 1000), the famed 1492 first voyage of Christopher Columbus was more truly the European "discovery" of America because Columbus publicized and followed up on his explorations. Although Columbus found little of material value in the New World, his four voyages triggered widespread European interest in exploration, exploitation, and colonization.

Spanish Settlement

In the wake of Columbus's first voyage, Pope Alexander VI divided the New World between Spain and Portugal. Although the provisions of this division (spelled out in the 1493 Treaty of Tordesillas) were generally ignored, the Spanish laid claim to much of North, Central, and South America, and the Portuguese claimed the vast territory of Brazil (after Pedro Cabral discovered it). While Spain and Portugal claimed vast portions of the New World, explorers sailing for other nations explored the Western Hemisphere not so much to exploit or colonize it as to find a way through it—a "Northwest Passage" that would be a shortcut to the already established trade markets of the East.

The Spanish explorers called themselves Conquistadors—conquerors—and acquired by force of arms the vast empire of New Spain. Cortez conquered the Aztec empire of Mexico, Pizarro the Incan empire of Peru, Ponce de León claimed Florida (and his countryman Menéndez founded St. Augustine there in 1565—a city now regarded as the oldest continuous settlement in North America), and Coronado and others explored the interior of what is now the United States. In the meantime, Balboa crossed the isthmus of Panama to reach the Pacific, and the crew of Ferdinand Magellan (who perished in the attempt) sailed around the world.

The Spanish searched in vain for riches—the fabled Seven Cities of Cibola or Seven Cities of Gold—and, in the process, established mines and plantations, exploiting Indians as slave labor. The Conquistadors were accompanied by Jesuit and Franciscan missionaries, who established a far-flung network of missions with the intent of converting the Indians to Christianity.

English Settlement

During the period of early Spanish exploration and conquest, King Henry VIII of England broke with Pope Clement VII over a variety of economic and political differences (including the pope's failure to sanction Henry's 1529 divorce and remarriage). This gave a powerful religious dimension to the growing rivalry between what was now non-Catholic England and Catholic Spain. Henry's daughter, Queen Elizabeth I, saw possession of territories in the New World as essential to countering Spanish power. She not only encouraged piracy against Spanish shipping, but sponsored such entrepreneurs and adventurers as Humphrey Gilbert and Sir Walter Raleigh in New World settlement schemes.

The first English attempts at colonization failed. Raleigh's colony in the land he called Virginia (in honor of Elizabeth, "the Virgin Queen") simply vanished without a trace and became known as the Lost Colony. In 1606, however, Elizabeth's successor, James I, granted charters to the London Company and the Plymouth Company, which financed the settling of Jamestown, Virginia, the next year. Disease and starvation nearly wiped the colony out, but its military leader, Captain John Smith, kept the enterprise from falling apart, and local Indians, led by Chief Powhatan, gave life-saving assistance. Among the lessons the Indians taught the colonists was the cultivation of tobacco, which rapidly emerged as a major export. Once cash began flowing in, the English colony expanded, and friction developed with the Indians, leading to open warfare. Indeed, relations between whites and Indians in America would be chronically and profoundly marked by violence through the end of the nineteenth century.

Virginia's agricultural economy grew quickly, resulting in an intense demand for labor. Well over half of the early settlers came to the colony as indentured servants, their passage expenses paid in return for their commitment to labor for their sponsor for a period (usually seven years). A headright system, whereby fifty acres was granted to settlers who paid their own way, was also instituted. Most profoundly and tragically, in 1619, the first shipment of African slaves—twenty in number—landed in Virginia. Slavery would become a way of life in the South. However, 1619 also saw Virginia institute a form of representative government when it convened its first General Assembly.

The Puritan Colonies

Just as it is important to recognize that the so-called New World was inhabited by peoples of differing cultures and backgrounds, so European settlement was motivated by various purposes. The Spanish sought gold, conquest, and glory for the Catholic Church. For the British of Virginia, American settlement was primarily a long-term commercial agricultural venture. For another group of

English settlers, however, religion was the prime mover.

Henry VIII had created the Church of England separate from the Catholic Church. Within the Church of England, a faction arose and grew, which sought further separation from the features of Catholic religious practice that remained a part of the Church of England. This group was known as the Separatists, and in 1609, suffering persecution from the Church of England majority at home, many migrated to religiously tolerant Holland. Yet they were English men and women, not Dutch, and they came to fear that raising their children in Holland would mean relinquishing their national identity. This motivated a move to America, under the auspices of the Virginia Company.

The first group of Separatists—who would come to be called the Pilgrims—sailed in *The Mayflower* and reached Cape Cod, in present-day Massachusetts, in 1620. Before finally debarking from *The Mayflower*, the leaders of the Pilgrim venture drew up an agreement known as the Mayflower Compact, which established certain democratic principles to govern the colony they called Plymouth, after their English hometown.

Like the settlers of Jamestown, the Plymouth colonists were poorly prepared for survival, and half the settlers perished in the first winter. Friendly Pokanoket Indians, led by Chief Massassoit (whose "ambassador" to the Pilgrims was Squanto), gave the newcomers aid and instructed them in the rudiments of New World agriculture.

The religion of the Separatists was known as Puritanism, a movement to return Christianity to its "pure" roots. A great wave of Puritan migration—from 1629 to 1640—followed the Pilgrim settlement of 1620. All settled in the region they called (and which is still called) New England. Puritanism became a powerful cultural force in America. It combined a sense of religious mission, a belief that the colonists were creating a "New Israel" (a new and absolutely righteous Holy Land) in the New World, with a drive to achieve economic success—what has become known as the "Puritan work ethic," a belief in the central importance and sanctity of hard work.

While Puritan society was intolerant of religious dissent, it also fostered important democratic principles. Each congregation, for example, was self-governing, and a General Court (legislature) was elected (albeit exclusively by male church members) to govern the colony. Nor was religious orthodoxy a matter of mindless obedience. There was a belief in individual conscience guided by strict interpretation of the Bible. In order to interpret the word of the Lord accurately, the Puritans founded Harvard College (and later Yale) to educate the clergy.

Although one would think that Puritan New England would be homogeneous and harmonious, it was, in fact, racked by religious dispute. Roger Williams made himself hateful to Puritan leaders by advocating a separation of church and state, by stressing the primacy of individual conscience over biblical interpretation, and by advocating respect for the rights of the Indians. Banished from Massachusetts in 1636, he founded the colony of Rhode Island, which became a haven of religious tolerance. It was to Rhode Island that Anne Hutchinson, a passionate advocate of religious inspiration through direct contact with the holy spirit (as opposed to the authority of the clergy), came.

A late example of the disintegration of the Puritan church was the witchcraft trials held in Salem, Massachusetts, during 1692. Twenty accused witches were executed and more than a hundred others were imprisoned.

The Middle Colonies

While New England was organized around a sharply focused religious and national theme, the group of colonies between New England and the Chesapeake Bay were founded by peoples of various religious and even national backgrounds.

The small country of Holland was an important trading power in the seventeenth century. It hired the English sailor Henry Hudson to search for a "Northwest Passage" through America and to the Far East. Such a passage would greatly expedite trade. Hudson failed to find the passage (it doesn't exist), but, in 1609, he did explore the river that now bears his name and, soon after this, the Dutch settled Manhattan (which they called New Amsterdam) and Albany (then called Fort Orange). From these outposts, Dutch settlers established a vigorous trade with the Indians in furs and other commodities. Although the Dutch grew powerful enough to absorb Swedish settlements along the Delaware River during the 1650s, Dutch colonial government was both intolerant and inefficient. Relations with the Indians broke down as well. Following British victory in the Anglo-Dutch Wars in Europe, England's King Charles II granted what is now the New York-New Jersey area to his brother James, Duke of York. In 1664, despite the efforts of the crusty Dutch governor Peter Stuyvesant to rally his people, the Dutch settlers surrendered their colony. New Holland became New York, and its chief town, New Amsterdam, was also called New York. That same year, the Duke of York established New Jersey.

Pennsylvania had a different origin. It was established as a refuge for persecuted Quakers (a radical religious group, who believed in direct inspiration from God, entirely without the authority of church or clergy) by William Penn, a well-connected Quaker. Penn made lasting treaties with the Indians and instituted the most democratic form of government among the early colonies. Because of its tolerance, Pennsylvania attracted a wide variety of immigrant groups (German religious groups being among the most prominent), and grew into a diverse and prosperous colony. Its chief city, Philadelphia, quickly outstripped New York as a commercial and cultural center and soon vied for preeminence with long-established Boston.

The Southern Colonies

The northernmost of the Southern colonies, Maryland was founded in 1634 by George Calvert, Lord Baltimore, a Roman Catholic who sought to create a refuge for other Catholics. The colony was by no means exclusively Roman Catholic, however; indeed, soon a majority of its population was Protestant, leading to the passage of the Act of Toleration (1649) to protect the rights of the Catholic minority. The act was one of the important early colonial documents of religious freedom.

As with the other Southern colonies, large-scale agriculture was the basis of the economy. In Maryland, the main export crop was tobacco, and while African slavery was practiced in this colony, most field labor was provided by white indentured servants. This was not the case in the Carolinas, where African slavery was soon very widely practiced. North and South Carolina were split from the original colony in 1729, with North Carolina attracting often poor settlers from Virginia, whose independent and democratic leanings were strong. South Carolina was dominated by planters from Barbados and the West Indies, who brought their slaves with them and established large rice and indigo plantations in the lowlands. Charleston developed as a sophisticated port city strongly influenced by the culture of the French Protestants who found safe haven there.

The last of the English mainland Southern colonies was Georgia, which was founded under a royal charter in 1732 by James Ogelthorpe. A visionary reformer, Oglethorpe sought to make his colony a practical utopia. It was populated in part by debtors for whom the colony was an alternative to debtor's prison in England, and the colonial government at first banned both slavery and alcohol. Both prohibitions were soon breached, and Georgia became a slaveholding plantation colony.

CHAPTER 7

ERA OF REVOLUTION (1763-1789)

MANY INTRODUCTORY-LEVEL AMERICAN HISTORY COURSES DEVOTE A CERTAIN amount of time to the rich cultural development of the seaboard colonial cities. Religion, literature, arts and crafts, newspapers, and education were all important. Great individual fortunes were made, and elaborate public buildings and mansion houses were constructed. Such displays of cultivation and wealth stood in stark contrast to the harsh subsistence-level conditions inland, away from the seaboard, on the frontier and in the wilderness. This contrast created many conflicts, as poor western settlers (correctly) felt that their interests were being served neither by the mother country nor by the eastern seats of government and economic power. In this regional and class conflict were the seeds of eventual revolution. Bacon's Rebellion (1676), a violent Virginia uprising that was quickly put down, was one early, dramatic result of this conflict.

Regional and class differences were hardly the only sources of early American conflict. As white settlement expanded, Indian wars became more frequent. In New England, the Pequot War (1637) resulted in the virtual destruction of that tribe, and King Philip's War (1675-75), between the New England colonies and a loose confederation of Indian groups under Metacomet (called King Philip by the English), was very destructive both to the colonists and to the Indians.

In the face of such conflicts, James II tightened his control of the American colonies, revoking royal charters and dissolving colonial assemblies. But when the Glorious Revolution (1689) overthrew James II, replacing him with William and Mary, colonists in Massachusetts, New York, and Maryland rebelled. These rebellions proved short-lived, as the British crown quickly reasserted control over its colonies, including very important restrictions on trade that benefitted the mother country largely at the expense of colonial mercantile interests.

The French and Indian Wars

If the colonies were often treated as economic pawns, they certainly figured as political pawns as well. The end of the seventeenth and much of the eighteenth century saw a series of dynastic wars in Europe, which expanded to global proportions in the nations' colonial possessions. The following European conflicts also had theaters of battle in North America. Collectively, the North American phases of these conflicts may be called the French and Indian Wars, because they involved British versus French colonials, with Indians usually fighting as French allies:

- **King William's War** (1689-97) was the North American phase of the War of the League of Augsburg, fought to checkmate French expansion. In North America, the French colonies—called New France—controlled important inland waterways and extended from the St. Lawrence River in Canada, through the Great Lakes, and down the Mississippi to the Gulf of Mexico at New Orleans. King William's War was a frontier conflict, involving raids on English settlements by French-allied Indians.
- **Queen Anne's War** (1701-13) was the North American phase of the War of the Spanish Succession, which was fought to block a union of France and Spain. Once again, French-allied Indians bore the brunt of the fighting against the English, with the conflict centered in New England as well as in the southern regions bordering Spanish Florida.
- **King George's War** (1744-48) was the American phase of the War of the Austrian Succession, fought between France and Spain on one side and England on the other. The fruitless power struggle in North America resulted in a restoration of the "status quo antebellum"—the way things were before the war.

The French and Indian War

These wars, fought simultaneously in North America and in Europe, were preludes to what many historians have called the "first world war," the Seven Years' War, which involved France and its allies and worldwide colonial possessions versus England and its allies and colonies. In North America, this great conflict was called the French and Indian War and was fought, mainly in the frontier regions, from 1754 to 1763.

The war began when the government of Virginia sent a young militia colonel, George Washington, to evict the French from western Pennsylvania. Washington suffered a defeat on July 4, 1754, and the following year British General William Braddock was disastrously defeated by the French at Ft. Duquesne (modern Pittsburgh). For the most part, until 1759, the British and the British colonials suffered defeat after defeat at the hands of the French and their Indian allies. This was due in part to the skill in wilderness battle demonstrated by the Indians and the French and was also the result of poor leadership among the British military and the failure of the colonies to unite effectively. But with the ascension of William Pitt to the post of English prime minister, the war effort found an adequate leader, and England's fortunes turned. English colonial forces, allied with the Iroquois (one of the few Indian groups that did not side with the French), retook territory in upstate New York, and British General Wolfe defeated French General Montcalm in the Battle of Quebec. Although both commanders perished in the battle, Quebec fell to the British. The war dragged on for another four years, but this defeat effectively spelled the end of French power in North America.

With the conclusion of the Treaty of Paris in 1763, ending both the Seven Years' War and the French and Indian War, England emerged as the dominant colonial power in North America and, indeed, the world. The nation now acquired all French territory in America, up to the Mississippi, as well as Spanish Florida.

Roots of Revolution

Even though the French and Indian War ultimately resulted in British control of North America, it also significantly weakened the mother country's hold on its colonies. The war had been a long and bloody one. During its first several years, British arms proved inefficient at best and disastrously unsuccessful at worst. Most of the fighting was on the frontier, and most of the British victories in those regions were due to the performance of colonial forces—sometimes *in spite of* poor leadership by British "regulars" (commanders of the regular British army). Increasingly, colonial leaders began to realize that they could not count on the mother country to aid in their defense and that, indeed, they were better off relying on their own resources. This attitude was especially prevalent in the outlying regions of the colonies, which were most vulnerable to Indian attack and which received the least military aid from British colonial authorities.

The growing discontent came to a focus when King George III of England issued the Proclamation of 1763 in an effort to placate still-hostile Indians (primarily those associated with Chief Pontiac) and to control the increasingly restive colonies. The proclamation drew a line along the Appalachian Mountains, forbidding white settlement west of the line. This outraged expansion-bent colonial frontiersmen, who largely ignored the proclamation. By violating the proclamation line, settlers provoked renewed Indian warfare, in which British authorities refused to intervene because the settlements were illegal. This further alienated the colonies from the mother country.

The Taxation Crisis

While the Proclamation of 1763 was driving a wedge between Britain and her colonies, the British government, led by finance minister Lord Grenville, intent on recouping the costs of the French and Indian Wars, levied a series of taxes and other restrictive measures on the colonies.

- **The Currency Act** (1764) barred the colonies from issuing their own currency.
- **The Sugar Act** (1764) levied duties on a variety of imports.
- **The Stamp Act** (1765) required a tax stamp on printed matter and legal documents.
- **The Quartering Act** (1765) required colonists to provision and house British troops.

These acts united the colonies in protest against "taxation without representation"—that is, against being taxed by the mother country in whose parliament the colonials were not directly represented. Various revolutionary societies—most prominently the Sons of Liberty—formed throughout the colonies, and a nine-colony Stamp Act Congress was convened in New York City in 1765. It drafted an effective protest to King George III, resulting in the repeal of the

Stamp Act in 1766 (although Britain affirmed Parliament's authority to govern the colonies in the Declaratory Act).

Repeal of the Stamp Act hardly brought the colonial crisis to an end. The Townshend Acts (1767) imposed a raft of new taxes and tariffs, triggering new protests, some violent, and prompting the dispatch of more British troops to Boston, now a hotbed of rebellion. A March 1770 confrontation between troops and a "patriot" mob resulted in the Boston Massacre when the troops fired into the mob. Among those killed was Crispus Attacks, an escaped slave; thus one of the first men killed in revolutionary activity was an African American.

While Lord North, the new British prime minister, repealed the Townshend duties, he left the tea tax in force. After a lull of about three years, Committees of Correspondence were formed to organize activity among American revolutionaries. In 1773 a new Tea Act was passed, which effectively forced colonial tea merchants out of business. In protest, a group of Bostonians, disguised as Indians, dumped a newly arrived cargo of British tea into Boston Harbor. Other "tea parties" occurred in Charleston, South Carolina, and Annapolis, Maryland.

Faced with escalating demonstrations of rebellion, Parliament passed the so-called "Intolerable Acts" in 1774:

- The port of Boston was declared closed until the tea was paid for.
- British government officials and soldiers were declared immune from trial by colonial courts.
- A new Quartering Act was passed.
- In Massachusetts, several elective offices were replaced by offices subject to royal appointment.

The Intolerable Acts prompted the First Continental Congress in September 1774, in which twelve colonies were represented. While the Congress agreed that Parliament might regulate external colonial commerce, it denied its authority over internal colonial affairs. A boycott of British imports was organized, and, in many places, official British governments were replaced by rebel-led committees of public safety. Although Lord North attempted conciliation, the effort collapsed, and Parliament declared Massachusetts to be in a state of rebellion.

The Revolution Begins

You don't have to be a student of history to know that we celebrate the anniversary of the 1776 Declaration of Independence each July 4th; however, the War for

Independence (the American Revolution) began some fifteen months before this date, in April 1775, with fighting in the Massachusetts towns of Lexington and Concord. British authorities had ordered the arrest of two prominent rebel leaders, John Hancock and Samuel Adams, as well as the seizure of munitions stores near Boston. Paul Revere and other members of the Sons of Liberty rode throughout the Boston-Concord-Lexington area to alert the colonial militia (so-called Minutemen, available for battle on a minute's notice) to resist the approaching British troops. Confrontation came at Lexington, where the first shots of the Revolutionary War were fired. An exchange at Concord took place next, and the British retreated to Boston, to which the colonial forces laid siege. Although the siege was broken in June by the British at the Battle of Bunker Hill (actually fought at nearby Breed's Hill), this victory came at a high cost for the British.

Revolution in the Northeast

While significant numbers of colonists remained loyal to the king (they were called Loyalists or Tories), the majority united in rebellion. In the face of unanswered petitions to King George III, Thomas Paine, a prominent revolutionary, wrote the eloquent pamphlet *Common Sense*, which crystallized the patriot cause and argued the necessity of creating an independent republic.

In May 1776, the Second Continental Congress convened and urged the colonies to form their own governments. A committee consisting of John Adams, Benjamin Franklin, and principal author Thomas Jefferson drafted the Declaration of Independence—in large part to justify the revolution to the world—which was adopted by the Congress on July 4, 1776.

Congress also appointed George Washington to lead the Continental Army. As for the British, they reinforced their army with more troops from England as well as with paid foreign mercenaries, most notably soldiers from the German state of Hesse (Hessians). While the British enjoyed the advantage of fielding a professional army, the rebels were fighting on territory familiar to them and typically employed the tactics of wilderness guerrilla warfare to good effect. One such guerrilla force, Ethan Allen's "Green Mountain Boys," captured Ft. Ticonderoga and its cannons in May 1775, but an early rebel invasion of Canada failed.

For the patriots, the low point of the war came during the New York and New Jersey campaigns, as the British captured New York City and pushed Washington's army through New Jersey and into Pennsylvania in December 1776. However, the day after Christmas, Washington recrossed the Delaware

River and scored a surprise-attack triumph at Trenton, New Jersey, and then, on January 3, 1777, at Princeton. Nevertheless, British forces took Philadelphia, and Washington wintered at Valley Forge, Pennsylvania, during 1777-78. His great task was to hold his poorly fed, poorly equipped army together during a most bitter winter.

As it turned out, the British capture of Philadelphia was not a wise strategic move. Britain's master plan was to capture the Hudson River, thereby cutting off New England from the rest of the colonies. British General Burgoyne succeeded in recapturing Ft. Ticonderoga, but British General Howe turned his attention to Philadelphia instead of coordinating the Hudson campaign with Burgoyne. In one of the war's most decisive battles, the Americans forced Burgoyne's surrender at Saratoga, New York, on October 17, 1777.

European Allies

Among the most important results of the Saratoga victory was that it persuaded France officially to join the war on the American side. The Marquis de Lafayette had been aiding Washington already, but now even more substantial French forces, especially naval forces, came to the aid of the Americans. Other foreign volunteers, including the Polish officers Kosciuszko and Pulaski and the Germans De Kalb and von Steuben, helped lead and train the colonial soldiers.

At Sea

At this time, Great Britain was the world's most formidable sea power, against which the fledgling American navy stood no chance. Britain used some of its fleet to blockade American ports and to supply its soldiers. However, American privateers (civilian raiders) effectively harassed British shipping, and Captain John Paul Jones boldly staged attacks on British warships in European waters. While Jones's actions had no great military consequence, they had a profound effect on patriot morale.

The entry of the French navy on the side of the Americans, along with the French-allied Spanish and Dutch fleets, had a much greater impact on British naval operation.

Revolution on the Frontier and in the South

The great battles of the eastern seaboard were important, of course, but much of the war was fought in small, sharp, bitter frontier skirmishes. This time, most of the Indians sided with the British, believing that a British victory would limit expansion of white settlements into their territory. While patriot settlers suffered greatly in the war, the Indians bore the brunt of the devastation, especially in the Ohio River country, where George Rogers Clark, together with French volunteers, captured several key British outposts. An important consequence of the war along the frontier was the opening of the wilderness region to American settlement.

For much of the war, the patriot forces had fared poorly in the South. Georgia fell to the British in 1778 and the key port city of Charleston, South Carolina, in 1780. In the Southern backcountry, however, American guerrilla commanders such as Nathanael Greene and Francis Marion (better known as the Swamp Fox) repeatedly foiled the British. Ultimately, it was a Southern battle, at Yorktown in October 1781, that brought about the final defeat of the British. Here British General Cornwallis surrendered to combined American and French forces, and while sporadic fighting continued for more than a year after Yorktown, the major battles were finished. At last in 1783, the British were ready to negotiate. The Peace of Paris (September 3, 1783), ended the war, and the British colonies were given their independence.

CHAPTER 8

THE EARLY REPUBLIC (1789-1824)

THE AMERICAN REVOLUTION WAS, OF COURSE, MOMENTOUS, ALTHOUGH ITS IMMEDIATE social effects were modest rather than sweeping. Anywhere from 20 to 30 percent of the colonial population had remained loyal to the king during the Revolution. It is a testament to the essential moderation of the independence movement that no mass reprisals were executed against the Tories; however, some Loyalist property was seized, and many Loyalists moved to British Canada. Since most Loyalists were from the wealthier classes, the net result was a limited degree of social leveling in the new United States.

Three groups received no benefit from the American victory. Women failed to achieve any measure of equal status with men. African-American slaves remained slaves, although the Revolution did mark an informal end to the institution in the North, where it was never widespread in any case. As for the Native Americans, as many of them had feared, their lot was worsened by the British defeat. Now no authority whatsoever stood in the way of the expansion of white settlement.

Articles of Confederation

It was one thing to make broad statements of political principle in the Declaration of Independence, but quite another to hammer out a document setting forth the specifics of a new republican government. In the thick of the

Revolution, in 1777, the Continental Congress adopted the Articles of Confederation, which were not ratified until 1781. The Articles did not so much create a nation as a "firm league of friendship" among the several states. There was, in effect, no central, federal government, but, rather, an assembly of individual states.

Under the Articles of Confederation, a great deal was accomplished, including the conduct of much of the Revolution and the successful negotiation of the Peace of Paris. After acrimonious dispute, the vast claims some individual states made to western lands were also resolved, and the disputed lands ceded to the nation "for the common benefit." The Land Ordinance of 1785 and the Northwest Ordinance of 1787 provided for the orderly organization of the vast Ohio country and the rational transition from territorial to state status. Importantly, the Northwest Ordinance provided a Bill of Rights to protect freedom of religion, the right to a jury trial, and other fundamental rights in the territory. Slavery was forbidden in the territory—and thus the Northwest Ordinance became the first federal stand against slavery.

Despite these achievements, the Articles of Confederation were fatally flawed. The federal government lacked the authority to levy taxes, to raise armies, to regulate commerce among the states, or to negotiate treaties with foreign powers. A financial debt crisis followed hard upon the Revolution, and the government was powerless to address it. Then, in 1786, Massachusetts farmer Daniel Shays led a march on the federal arsenal at Springfield, Massachusetts, in an attempt to foment an armed rebellion to halt foreclosure on the property of heavily indebted farmers. The federal government was powerless against Shays, and it was a state militia force that finally crushed the rebellion.

Constitutional Convention

The Shays affair was only the most dramatic of the many demonstrations of the inadequacy of government under the Articles of Confederation. In the summer of 1787, a convention was called in Philadelphia for the purpose of revising the Articles. Ultimately, revision was rejected in favor of creating an entirely new Constitution. It provided for:

- **A "bicameral" (two-chambered) legislature:** a Senate, in which all states would have equal representation, regardless of population, and a House of Representatives, in which seats were allotted according to state population

◆ **A truly federal system:** power and authority were shared by national and state governments

◆ **Separation of power:** A so-called "system of checks and balances" became the cornerstone of government structure. The powers of the executive branch, the legislative branch, and the judicial branch were structured such that no branch could become more powerful than another. By this it was hoped that tyranny and other excesses could be avoided.

The new Constitution failed to outlaw slavery. It addressed the issue chiefly as it related to representation and taxation, apportioning each according to a formula whereby a slave was to be counted as three-fifths of a person.

Bill of Rights and Ratification

Ratification of the new constitution did not come easily, and during the ratification period Alexander Hamilton, James Madison, and John Jay, writing under the collective pseudonym of Publius, issued a series of 85 *Federalist* essays, masterful arguments in favor of the new constitution. One of the key objections to the constitution, the absence of a bill of rights, was met by the addition of the first ten constitutional amendments, collectively called the Bill of Rights (1791), guaranteeing both personal (including religious) freedoms and stressing certain powers reserved not to the federal government but to the states.

Government under Washington and Hamilton

In the first presidential election in the new country (1789), George Washington was unopposed. Indeed, the framers of the Constitution felt confidence in granting considerable powers to the chief executive (the president) precisely because they envisioned Washington's election. Not only had this man shown himself to be a great military leader during the Revolution, he exhibited a conservative nobility of character 180 degrees removed from tyranny. The framers had no reason to fear that the first president of the United States would be a dictator.

Their confidence was well placed. The importance of the Washington presidency cannot be overstated. Not only did he lead the nation through the first years of its infancy, he did nothing less than set the pattern for the office he occupied. During his two terms as president, Washington:

◆ refused such noble titles as "Your Highness" and "Your Excellency," insisting that the president be addressed simply as "Mr. President"

- negotiated important treaties with England and Spain
- created a national bank
- established a policy of U.S. neutrality in the ongoing warfare between England and France (this created a strong tradition of U.S. isolationism and avoidance of "foreign entanglements")
- quelled a frontier Pennsylvania tax revolt known as the Whiskey Rebellion—thereby asserting the authority of the federal government in a way that had been impossible under the Articles of Confederation
- established the principle of executive privilege when he refused to submit to the Senate documents relating to an Indian treaty; thus Washington defined the limits of the constitutional provision of the Senate's role to "advise and consent" in such executive matters as treaty negotiation
- declined a third term, setting a precedent that was not broken until the four-time election of Franklin Roosevelt (The 22nd Amendment now legally limits the president to two consecutive terms.)

Under Washington, the presidency became a strong but non-dictatorial office of great dignity.

Besides Washington, another member of his administration had a profound effect on the new nation. Alexander Hamilton, Washington's secretary of the Treasury, created a strong financial plan that not only consolidated federal power, but did much to promote a strong mercantile and industrial economy. Moreover, Hamilton successfully recommended that all foreign debts (incurred during the Revolution) be paid, thereby establishing good credit for the nation. Hamilton's program was controversial, and it was the excise tax on whiskey he proposed (and Congress approved) that triggered the whiskey rebellion, a serious early challenge to federal power and authority.

Sound finances were not the only source of early federal credibility. When the French foreign minister to the United States, Citizen Edmond Genêt, plotted with U.S. privateers (civilian ship captains) to prey on English shipping in U.S. waters and, without executive approval, recruited support for the French Revolution, Washington successfully asserted national sovereignty by securing his recall to France. (In fact, Genêt ultimately became an American citizen to avoid returning to a nation now controlled by factions hostile to him.) Internally, Washington sent General "Mad Anthony" Wayne to secure the Northwest Territory (the Ohio frontier) from Indian hostility. His victory at the Battle of Fallen Timbers (1794) led to the Treaty of Greenville (1795), which brought under U.S. control the vast western territory north of the Ohio River.

Birth of the Party System

Washington was firmly opposed to "political factions" and political parties, but two more-or-less opposing parties developed even during Washington's tenure in office. The Federalists favored a strong central government, encouraged the growth of commerce and manufacturing, and wished to develop strong ties with England. The Republicans (not to be confused with the modern party of that name) favored states' rights and the agricultural way of life. The Republican affinity was with radical France rather than conservative England. While the Federalists stressed order and stability, the Republicans emphasized power in the hands of the people.

Government Under Adams and Marshall

The 1796 presidential election put John Adams in office with Thomas Jefferson (runner up in the election) as vice president. Adams immediately faced a crisis with France, which had effectively begun an undeclared naval war against the United States by seizing American shipping. When the envoys Adams sent to Paris were met by three agents of French Prime Minister Talleyrand (referred to as X, Y, and Z by Adams) with demands for bribes, Congress and the American people were outraged and embarked upon a major naval shipbuilding effort. Adams resisted war fever, however, and successfully negotiated an agreement with France.

The year 1798 also saw the Federalist Congress pass the Alien and Sedition Acts, repressive legislation intended to curb Republican criticism of the Federalist-dominated government. In response, Vice President Jefferson and James Madison drew up resolutions setting forth the states' right to nullify (in effect refuse to obey) any federal law the state considered unconstitutional. The Alien and Sedition Acts would largely be overturned during the Jefferson administration, but the nullification theory would continue to haunt American political life for more than fifty years, ultimately helping to spark the Civil War.

In a revolt against Federalist conservatism, Republican Thomas Jefferson was elected to the presidency in 1800. Before leaving office in 1801, Adams, however, made a key Federalist appointment to the Supreme Court when he elevated the brilliant John Marshall to chief justice. During some thirty-five years presiding over the high court, Marshall wrote landmark decisions that permanently shaped the role and authority of the Supreme Court. In *Marbury v. Madison* (1803), Marshall established the principle of judicial review, by which the Supreme Court became the arbiter of the Constitution and of the constitutionality of legislation. In *McCulloch v. Maryland* (1819), the Marshall court established the precedence of federal over state laws in cases where the two conflicted. In *Gibbons v. Ogden* (1824), the authority of Congress to regulate interstate commerce was affirmed.

The Age of Jefferson

Jefferson took office declaring his intention to reconcile Federalist and Republican differences, but in fact introduced a new Republicanism into American government by allowing the Alien and Sedition Acts to lapse, by repealing certain federal taxes, by reducing government spending, and by reducing the size of the army and navy. Nevertheless, when the so-called Barbary Coast pirates harassed American shipping off the coast of North Africa, Jefferson acted swiftly by dispatching a naval squadron, which forced the Pasha of Tripoli to conclude a peace treaty in 1805.

It is important to note that the transition from Adams and Federalism to Jefferson and Republicanism was profound, but also peaceful, orderly, and lawful, demonstrating the essential soundness of American constitutional government.

Louisiana Purchase

Jefferson's single most popular and sweeping achievement was the Louisiana Purchase of 1803, by which the United States acquired control of the Mississippi River, the gulf city of New Orleans, and an amount of territory that doubled the size of the United States. The purchase laid the groundwork for the United States' eventual claim on territory stretching from sea to sea. The brilliant and scientifically curious Jefferson commissioned two army officers, Meriwether Lewis and William Clark, to explore the extent of the Louisiana Purchase and beyond. This was a first step in the opening to settlement of the Far West.

War of 1812

Jefferson ended his first term on a note of triumph, but his second term was plagued by the disintegration of relations between the United States and England. Jefferson's object was to keep the nation neutral in the great conflict between France and Britain—the Napoleonic Wars—but when Britain blockaded the North American continent (in an effort to disrupt trade with France), then seized American shipping and, worse, boarded U.S. vessels to *impress* seamen, American popular anger was roused. (*Impressment* was the act of boarding a ship at sea, then forcibly seizing any sailors judged to be British subjects in order to "press" them into service in the Royal Navy. The U.S. regarded it as a gross breach of national sovereignty and an act of war.)

In response to British aggression, Jefferson initiated passage of the Non-Importation Act (1806) and the Embargo Act (1807), which effectively suspended trade with Britain. Unfortunately, these acts affected England far less than they damaged New England commerce and Western farming operations. The embargo was repealed just before Jefferson left office in 1809.

By the time James Madison entered the White House as the nation's fourth president, "War Hawks" in Congress (Henry Clay, John C. Calhoun, and others) were successfully agitating for war with England on the grounds of:

- impressment of seamen
- British provocation of and aid to Indian attacks on frontier settlers
- conquest (the most extreme War Hawks advocated invasion and conquest of Canada)

When President Madison at last asked Congress for a declaration of war (June 1, 1812), he cited the first two grounds only (though, on the eve of war, Britain actually agreed to cease impressment, and the British role in Indian provocation was never entirely clear and almost certainly not officially sanctioned). It was congressional support from Southern and Western War Hawks that secured passage of the declaration of war.

The War of 1812 was, for the most part, a U.S. disaster. Attempts to invade Canada failed, most frontier engagements went to the British and their Indian allies (so that the West was threatened with conquest rather than expansion), and a British invasion of the eastern seaboard began with the burning of many public buildings in Washington, D.C. Nevertheless, American morale and nationalist pride were bolstered by isolated but dramatic naval victories on the Atlantic and a spectacular victory on the Great Lakes by Oliver Hazard Perry. After hastily building and assembling a fleet on Lake Erie, Perry, on September 10, 1813, defeated the British lake fleet, cutting off British supply lines and enabling General William Henry Harrison to defeat the British (and their Indian allies under the great warrior-leader Tecumseh) at the Battle of the Thames (October 5).

In the meantime, in the South, General Andrew Jackson defeated the British-allied Creek Indians at the Battle of Horseshoe Bend (March 29, 1814), then went on to triumph over the British at the Battle of New Orleans (January 8, 1815).

The victory at New Orleans made a national hero of Jackson and created among Americans the feeling that they had achieved a glorious victory in the war, when, in fact, the Treaty of Ghent (signed in the Belgian city on December 24, 1814—before the momentous Battle of New Orleans) did nothing more than restore the "status quo ante bellum" (the state of things *before* the war). Nevertheless, the War of 1812 had important effects on the nation:

- It simultaneously aroused sectionalism—with New England and the Middle Atlantic states opposing the conflict, and the western and southern states favoring it—and national pride (the war has sometimes been called the "Second War of Independence").
- It significantly weakened the Indian tribes of the South and the Ohio country.
- It made heroes of two future presidents, Andrew Jackson and William Henry Harrison.

In the afterglow of the war, the U.S. Navy successfully renewed action against the Barbary Pirates. Domestically, Congressman Henry Clay introduced the "American System," which included a protective tariff to encourage U.S. industry, a rechartered (second) national bank, and funding for a national transportation system to open up the West. (President Madison vetoed this last provision, however.)

The "Era of Good Feeling"

The administration of President James Monroe (1817-25) was popularly known as the "Era of Good Feeling," despite a severe depression (an after-effect of the War of 1812) and sectional strife (which would intensify throughout the century as the nation inched toward eventual civil war).

The economic crisis was triggered by the Panic of 1819, which ushered in three years of severe depression caused by the failure of overextended banks (which had made bad loans to western land speculators) and the falling price of cotton and other agricultural goods. The sectional crisis deepened as the addition of new territories and states changed the balance between slave states and free.

In 1820, Henry Clay introduced the Missouri Compromise to cope with the crisis created by Missouri's application for entry into the union as a slave state. The compromise prohibited slavery in the Louisiana territory north of a 36' 30'—except in the case of Missouri, north of this line, which would be admitted as a slave state. To preserve the balance between slave and free states, Maine was separated from Massachusetts and admitted as a free state.

Despite domestic turmoil, the Monroe administration created a bold presence in international affairs. The United States recognized the sovereignty of newly independent Latin American nations during 1808-22 and, in his annual message to Congress on December 2, 1823, the president promulgated what came to be called the Monroe Doctrine, a declaration to the world that the United States would tolerate no new European colonization in the Western Hemisphere, but that the nation would not intervene in European affairs or in the affairs of existing European colonies. Great Britain and other European powers did not take the Monroe Doctrine seriously, since the United States lacked the military muscle to back up the policy; however, it evolved throughout the course of American history as the centerpiece of U.S. foreign policy.

CHAPTER 9

THE AGE OF JACKSON AND WESTWARD EXPANSION (1824-1850)

THE ELECTION IN 1828 OF ANDREW JACKSON, THE "HERO OF NEW ORLEANS," AS PRESIDENT signified a new era in United States politics. Previous chief executives had been members of the "Tidewater aristocracy," easterners, whereas Jackson was a "westerner"—that is, a man from the nation's frontier region. His election ushered in the "age of the common man" and was deemed by many a new era in democracy.

In a bitter contest, Jackson (member of the party now called the Democratic Republicans) defeated John Quincy Adams (running for reelection as a National Republican). Jackson introduced a civil service system that was soon criticized as a "spoils system," in which loyal party members were rewarded with patronage jobs in the government. Although Jackson favored a limited central government, he greatly strengthened the office of president, making extensive use of his veto power to defeat congressional bills he disliked. His supporters saw him as a strong champion of the "little man," while critics derided him as "King Andrew."

Jackson vs. The Bank of the United States

Jackson's favorite target was the Second Bank of the United States, which he saw as a financial monopoly favoring wealthy easterners and making it difficult for western farmers to obtain credit. Jackson removed federal funds from the bank and deposited them in state-chartered ("pet") banks, which printed inflationary paper currency. The result was easier credit for western expansion, but also financial instability that soon brought on an economic panic and depression.

Nullification Crisis

In 1816, Congress had enacted a protective tariff to encourage American industry. Tariffs were raised again in 1824 and in 1828. While these import duties aided the industrializing Northeast, they penalized the agricultural South and frontier regions by making it difficult for them to export raw materials such as cotton. John C. Calhoun led a fight to nullify the 1828 "Tariff of Abominations" (as its critics called it), arguing (much as Jefferson had done years earlier in protesting the Alien and Sedition Acts) that an individual state had the right to declare null and void any federal laws it deemed unconstitutional. Calhoun failed to muster sufficient support in the South Carolina state legislature to carry the nullification resolution.

Under President Jackson, a new tariff was enacted in 1832; while lower than the Tariff of Abominations, it was still protective, and, this time, a special South Carolina convention did nullify it. The state even threatened to secede from the union. To the shock and dismay of Calhoun (now Jackson's vice president) and many other southerners, Jackson responded with accusations of treason and threatened to send a federal military force to collect tariffs. At this, support for nullification from other Southern states dissolved. A new compromise tariff was enacted in any case, and South Carolina withdrew its nullification ordinance. The crisis subsided—but was a portent of ultimate disunion, later in the century, in the form of civil war.

Indian Removal

Among Thomas Jefferson's motives for making the Louisiana Purchase was the acquisition of a vast territory to which the Indian tribes of the East could be peacefully "removed." In 1824, President Monroe urged tribes to move west of the Mississippi voluntarily, but to no avail. At last, in 1830, with President Jackson's strong endorsement, Congress passed the Indian Removal Act, which provided for the compulsory resettlement of the eastern tribes in western lands reserved for them.

In some ways, the Indian Removal Act attempted to be fair —for example, the removed tribes were to be justly compensated for the lands they relinquished—but many tribes did not want to move. Warfare with the Black Hawks in Illinois

(1832) and with the Seminoles of Florida (beginning in the 1830s) resulted. Even more tragic was the removal of the Cherokee from Georgia and the Carolinas along a forced-march route to "Indian Territory" (mostly present-day Oklahoma), which became known as the Trail of Tears after many of the Indians succumbed to the effects of disease, bad weather, and what can only be described as broken hearts.

Age of Reform

Andrew Jackson handpicked his successor, Martin Van Buren, who ably coped with the Panic of 1837 by developing a federal subtreasury system that was independent of unreliable state-chartered and private banks. In many other ways, too, Van Buren was an able chief executive, but he was nevertheless defeated for reelection in 1840 by War of 1812 hero William Henry Harrison, the candidate of a new political party, the Whigs. This party opposed what it called "Jacksonian tyranny" and reintroduced the idea of a stronger central government, protective measures to encourage economic development, and general humanitarian reform. Harrison, however, died within a month after assuming office, and his vice president, John Tyler, served out his term. A Democrat, James K. Polk, was elected in 1844.

Reform had been in the air since early in the century:

- ◆ The Second Great Awakening (the first had occurred in the in the mid-eighteenth century) was a national evangelical revivalist movement.
- ◆ Among the many religious sects that arose early in the nineteenth century, the most enduring was the Church of Jesus Christ of Latter-Day Saints, the Mormons, which originated in upstate New York in 1830. In 1847, Mormon leader Brigham Young would lead members of the sect to a settlement on the Great Salt Lake in Utah Territory. Although the Mormon practice of polygamy caused much controversy (some of it violent), the Mormons played an important role in opening the far West to settlement.
- ◆ Utopian movements abounded in the first half of the century, including the Shakers, the Oneida community, the Owenites, and others.
- ◆ The atmosphere of radical reform gave rise to a temperance movement, as well as to movements promoting "free love," world peace, and humane treatment of criminal prisoners and the insane.

◆ Women's rights became a key issue. In 1848, feminists Lucretia Mott and Elizabeth Cady Stanton called the Seneca Falls (N.Y.) Convention and issued a Women's Declaration of Independence. Susan B. Anthony emerged as an important advocate of women's rights, including the right to vote.

Slavery and Abolitionism

At this point in American history, from the 1830s to the Civil War, no reform movement was more important, vocal, or consequential than abolitionism—the fight to end slavery. While the importation of slaves had been outlawed in the United States in 1807, slavery continued as a way of life in the South and was felt to be crucial to the Southern economy. Various abolitionist movements developed, even as early as the eighteenth century, but it was William Lloyd Garrison's radical abolitionist newspaper, *The Liberator*, that first rallied truly organized and widespread support for the movement beginning in 1831.

Abolitionists worked to end slavery by legal as well as extralegal means. The Underground Railroad was a system of volunteer whites and free blacks dedicated to aiding slaves in their flight from the South to the North. Slavery and opposition to it progressively tore the nation apart, leading inexorably to civil war.

Texas Independence

As a result of the Louisiana Purchase, U.S. territory already extended far west of the Mississippi River. In 1820, Moses Austin secured a grant form the Spanish government of Mexico to found an American colony in Texas, at the time a Mexican state. Moses Austin died, leaving his son, Stephen, to carry out his plans. By the mid-1830s, Americans far outnumbered Mexicans in Texas, and in 1836 the Texans began a war of rebellion to gain independence from Mexico. When Mexican general Santa Anna, leading an army of 5,000, attacked and ultimately massacred the 187 Texans (led by William B. Travis) defending an improvised fortress known as the Alamo, the Texan War for Independence drew national attention and support. On April 21, 1836, Sam Houston brilliantly led a ragtag army to victory against Santa Anna at the Battle of San Jacinto, and Texas became an independent republic.

Most Texans wanted the United States to annex their new nation, so that it could eventually become a state. It was clear, however, that Texas sought entry into the union as a slave state, which was opposed by the legislators representing the free states. Moreover, it was realized that annexation would provoke war with Mexico, something few U.S. politicians were eager for at this point. The issue of Texas annexation and statehood was, therefore, tabled.

Manifest Destiny and a War with Mexico

Despite hesitancy to annex Texas, the national drive to continue expanding settlement westward was strong. In 1845, *New York Post* editor John L. O'Sullivan wrote that "It is our manifest destiny to overspread and possess the whole of the continent which Providence has given us for the development of the great experiment of liberty and federated self-government entrusted to us." That phrase "manifest destiny" stuck, becoming a watchword for conquering the hard soil of the western prairies, the harsh climate of the plains, and the hostility of western Indians (some having been removed from the East, some native to the region).

At last, outgoing president John Tyler urged Congress to adopt a resolution to annex Texas, and in 1845, Tyler's successor, James K. Polk, signed the resolution. Texas immediately became a state, and this immediately triggered a border dispute with Mexico.

Polk next turned to California, a Mexican possession. He offered to purchase the territory for $40 million, but his emissaries were rebuffed. The dispute over the Texas border and the possession of California quickly erupted into the Mexican-American War.

In the South and West, the war was wildly popular; New Englanders, however, resisted it, protesting that it was an unjust war of conquest. Such protests notwithstanding, American forces not only readily took California, but invaded Mexico and marched into its capital city. The Mexican-American War ended with the Treaty of Guadalupe Hidalgo (ratified by the Senate on March 10, 1848), by which Mexico ceded to the United States New Mexico (which included parts of the present states of Utah, Nevada, Arizona, and Colorado) and California. The Mexicans further renounced claims to Texas above the Rio Grande.

Immigrant Trails and a Rush for Gold

By the time of the Mexican-American War, the vast West was being pierced by a rudimentary system of trails along which organized wagon trains moved "immigrants" (as the settlers were called) to new homes on the plains, in California and the Oregon Territory, and in the Southwest.

The lure of open land was a mighty magnet for mass movement, but an even stronger force suddenly emerged in 1848. On January 24 of that year, James Wilson Marshall, in the employ of Northern California rancher Johann Augustus Sutter, discovered gold in the race (water course) of a mill on Sutter's property. Within a year, numerous gold discoveries were being made throughout the region, and the nation fell into the grip of gold fever. Tens of thousands of easterners left their jobs, literally dropping their tools where they stood, and lit off for the California gold fields. While few succeeded in finding their fortune, the Gold Rush helped to populate much of California and, later, Nevada, Colorado, and the Dakotas.

Westward migration, whether from motives of agricultural settlement or gold prospecting, strengthened the bonds joining East and West, even as those joining North and South steadily dissolved over the issues of slavery and abolition.

CHAPTER 10

THE ERA OF CIVIL WAR AND RECONSTRUCTION (1850-1877)

THE MISSOURI COMPROMISE OF 1820 STAVED OFF CIVIL WAR, BUT NO ONE WAS EVER very happy with it. Thomas Jefferson said that, "like a fire bell in the night, [the Compromise] awakened and filled me with terror," and John Quincy Adams called it the "title page to a great tragic volume." Sure enough, abolitionist and slave-holding factions continued to battle, and each new territorial acquisition and application for statehood created a crisis in the precarious balance of free versus slave states.

Compromise of 1850

California's application for entry into the union as a free state following the war with Mexico created the usual crisis. In response, five bills, collectively called the Compromise of 1850, were passed by Congress, providing for:

- entry of California as a free state
- creation of New Mexico and Utah territories—with the issue of slavery here to be determined by *popular sovereignty* (that is, by a vote by the citizens of the territory, not federal legislation)
- abolition of the slave trade in Washington, D.C.
- enactment of a strong Fugitive Slave Law

The latter provision, which made it much more difficult for sympathetic northerners to aid and harbor runaway slaves, created the most dissension. Some Northern states passed laws contrary to the federal law, and, in some places, the return of fugitive slaves was resisted with violence. If anything, the new Fugitive Slave Law galvanized the abolition movement and increased the activity of the Underground Railroad.

"Bleeding Kansas"

By the end of the 1840s, projects of building a great transcontinental railroad were much discussed. Partly to promote the choice of a northern route for the railroad, Senator Stephen A. Douglas of Illinois introduced the Kansas-Nebraska Bill (1854) to create those two territories. The bill provided for determination of the slavery question by popular sovereignty, thereby erasing the demarcation line drawn by the Missouri Compromise. Nebraska, a Northern territory, would surely vote itself free; however, Kansas was up for grabs, and anti-slavery ("free-soil") northerners and pro-slavery Missourians alike rushed into the territory to vote. The Missourians moved faster, and the territorial legislature was occupied by a pro-slavery majority, which not only voted the territory open to slavery, but created a strict slave code to ensure that no fugitives would be harbored. In response, the free-soilers drew up the rival Topeka Constitution, outlawing slavery. A period of intense guerrilla warfare ensued in what came to be called "Bleeding Kansas."

Birth of the Republican Party

The stress on the nation created by the slavery issue was reflected in the disintegration of old political parties and the creation of new ones. The most important party to emerge was the Republican Party, formed in 1854 as a coalition of various parties united in their opposition to slavery—or, more precisely, their support of free labor and their opposition to extending slavery into the territories. (That is, the Republicans were not radical abolitionists favoring an immediate end to slavery.)

Although John C. Frémont, the party's first presidential candidate, lost to Democrat James Buchanan, the Republican Party rapidly grew in strength and soon attracted an Illinois politician named Abraham Lincoln.

The Dred Scott Decision

In 1857, the slavery controversy suddenly shifted from the halls of Congress to the bench of the Supreme Court. Dred Scott was a slave whose owner, an army surgeon, had taken him into posts in a free state and in a free territory. Based on his residence in these areas, Scott (backed by abolitionist forces) sued for his freedom. The courts of Missouri (a slave state) repeatedly ruled against him, but ultimately the case was heard by the U.S. Supreme Court. Chief Justice Roger Taney wrote for the 6-3 majority:

- Scott, a slave, was not a citizen, and therefore could not bring suit.
- In any case, his residence in a free state and free territory did not confer freedom on him once he returned to Missouri.
- The Missouri Compromise, which had prohibited slavery in a territory, was declared unconstitutional.

The Dred Scott Decision polarized public opinion. In Illinois, it became the subject of a brilliant (and nationally publicized) series of debates between senatorial candidates Abraham Lincoln and Stephen A. Douglas. Although Lincoln lost the election, he gained national prominence. In any event, Lincoln and Douglas were not all that far apart on the slavery issue. Both opposed slavery, but Douglas favored a hands-off approach by the federal government, proposing instead the so-called Freeport Doctrine, which suggested that anti-slavery territories did not have to outlaw slavery explicitly, but could simply refuse to enact the slave codes that protected the institution.

The Freeport Doctrine and other such temporizing positions could not long contain the opposing forces aligned on either side of the slavery issue. In 1859, John Brown, a radical abolitionist who had made a name for himself when he led a murderous attack on pro-slavery forces at Pottawatomie Creek, Kansas (May 24, 1856), occupied the federal arsenal at Harpers Ferry, Virginia (today West Virginia), with the intention of arming slaves for a mass rebellion. Brown was captured, tried for treason, convicted, and executed; however, he and his deed became a rallying point for abolitionists. Many regarded Brown as a martyr to the cause.

Emergence and Election of Lincoln

In an atmosphere of violence and impending violence, Abraham Lincoln secured the Republican nomination and, in 1860, won election as president by garnering a plurality (40 percent of the vote) in a field of candidates whose constituents were badly split. Although Lincoln and his party merely opposed the *extension* of slavery, South Carolina, citing the election of a president "hostile to slavery," seceded from the union on December 20, 1860.

Secession Crisis

Within months, ten other Southern states followed South Carolina's lead, and the Confederacy was born. Senator John Crittenden, of Kentucky, a "border state" (a state with slavery and Southern ties, but which did not secede), proposed a compromise constitutional amendment that would protect slavery where it existed south of the original Missouri Compromise line. A Peace Conference (February 1861) was held in Washington. Ohio Senator Thomas Corwin proposed an amendment protecting slavery *wherever* it presently existed. None of these measures averted the coming storm.

Civil War: Fort Sumter to Antietam

At 4:30 a.m. on April 12, 1861, South Carolina cannons began firing on Fort Sumter, in Charleston Harbor. At length, the fort's commandant, Major Robert Anderson, his ammunition exhausted, surrendered Sumter, and the Civil War began.

Although the population and economic and industrial might of the North far outweighed the technology and numbers of men the South could muster, the cream of the U.S. Army officer corps felt allegiance to the Southern states, and the Confederate forces were, in the main, more ably commanded than those of the North, especially early in the war. This fact was brought home by the Confederate victories at Bull Run (Manassas; July 21, 1861), the so-called Seven Days (during the Peninsular Campaign of Union commander George B. McClellan), the Second Battle of Bull Run (Second Manassas; August 29-30, 1862), Fredericksburg (December 13, 1862), and Chancellorsville (May 2-4, 1863). At last, Antietam (Sharpsburg), on September 17, 1862, brought Union forces a very costly, very narrow victory.

Civil War: The Emancipation Proclamation

The outcome of Antietam gave the Union military effort sufficient credibility to encourage President Lincoln to issue a preliminary Emancipation Proclamation, which, officially bringing the slavery issue into the center of the conflict, conferred on the Civil War an added moral dimension as a crusade to abolish slavery. A "final" Emancipation Proclamation was issued on January 1,

1863, which liberated only those slaves in areas "still in rebellion"; that is, in the Confederate areas presently under Union control and in the border states, slavery continued. (The Thirteenth Amendment, ratified by the states on December 18, 1865, after the war was over, made abolition the law of the land.)

Civil War: Gettysburg to Appomattox

By 1863, the South's great general, Robert E. Lee, adopted an offensive strategy, hoping that, by bringing the war to the North, that side would be willing to negotiate a favorable peace. Union forces under General George G. Meade turned back Lee's army at Gettysburg, Pennsylvania (July 1-3, 1863), a battle usually cited as the turning point of the war in favor of the Union. Not only was an invading army repulsed and Northern morale raised, the Southern defeat discouraged England and France, both potential Confederate allies, from rallying to the Southern cause.

While most of the fighting was concentrated in the East, the war's western theater was also highly important. Union control of the lower Mississippi began with the victory of Admiral David Farragut, who captured New Orleans in April 1862, then General Ulysses S. Grant scored major victories at Shiloh (April 6-7, 1862), Vicksburg (under siege from October 1862 to July 4, 1863), and Chattanooga (November 23-25, 1863). The following year, Lincoln (who had gone through a series of mediocre commanders) appointed Grant as the Union's general-in-chief. Grant slowly pushed Lee's army back toward the Confederate capital of Richmond, Virginia. Grant' s chief lieutenant, General William Tecumseh Sherman, simultaneously advanced through Tennessee and Georgia to Atlanta, which he captured, occupied, and finally burned (September-November 1864) before continuing on his infamously destructive "march to the sea."

Despite the demoralizing effect of Sherman's wholesale destruction of the civilian South, the fighting continued. Only after a long campaign did Grant take heavily fortified Petersburg, Virginia, which put him in position to capture Richmond on April 2, 1865. A week later, at Appomattox Courthouse, General Robert E. Lee surrendered his Army of Northern Virginia to General Grant, effectively ending the Civil War.

Assassination of Lincoln

No president had a more wrenching tenure than Abraham Lincoln, presiding over a torn nation engaged in slaughtering its own. Yet, in the inaugural address marking his election to a second term, Lincoln spoke not just of the coming victory, but of reconciliation, of healing, of charity for all and malice toward none. Healing the wounds of war and reuniting the nation were staggering tasks that would require the wise and charitable judgment of no less a figure than Lincoln. His assassination by a Southern sympathizer, the actor John Wilkes Booth, on April 14, 1865, was the crowning tragedy of the Civil War. Booth's self-proclaimed act of vengeance on behalf of the Confederacy deprived the South of a chief executive who would have treated the region with moderation and generosity—and who would have stood the best chance of persuading his Northern constituents to reconciliation.

Instead, the unpopular Andrew Johnson was thrust into office. Johnson was a senator from Tennessee, who remained in the Senate after his state seceded. After Tennessee was recaptured by the Union, Johnson served as military governor and, in 1864, was chosen by Lincoln as his vice presidential running mate in a gesture of national unity. Although Johnson's approach to Reconstruction (as the process of reunifying the nation after the Civil War was called) was similar to the slain Lincoln's, Johnson lacked Lincoln's heroic character and powers of persuasion. He clashed with the Radical Republicans in Congress (who were bent on punishing the South—and delaying the return of Southern Democrats to Congress), vetoing Reconstruction legislation and often refusing to enforce the legislation passed over his veto. At last, when Johnson attempted to remove Secretary of War Edwin Stanton from the cabinet, Congress impeached him on the grounds of having violated the Tenure of Office Act (which required Senate approval for removal of presidential appointees).

The impeachment was politically motivated, and despite an overwhelming Radical Republican majority in the Senate, Johnson was saved from removal from office by a single vote. Johnson served out his term, powerless to control or curb the punitive Reconstruction policies of the radicals.

Reconstruction: Civil Rights and Civil Wrongs

Most of the Reconstruction period was presided over by the two-term administration of Ulysses Grant—who had been a great general, but proved an inept president. While personally honorable, Grant suffered through an administration riddled with an unprecedented level of corruption and scandal and plagued by postwar economic depression.

Many of the policies of Reconstruction were well-intentioned:

- ◆ Civil rights protection for blacks was put in place.
- ◆ Universal manhood suffrage was established.
- ◆ Criminal codes were reformed.
- ◆ Certain economic recovery programs were created.

In practice, however, Reconstruction fell far short of its ideals. Recovery programs were administered both ineptly and corruptly. Carpetbaggers—Northerners who "invaded" the South after the war—exploited freedmen (former slaves) in order to control local governments and profit from graft and corruption.

Partly in response to the hated policies of radical Reconstruction and partly out of long-ingrained racism, many white southerners lashed out at blacks. Militant secret societies, especially the Ku Klux Klan, terrorized blacks (as well as white Republicans) throughout the South. Beatings, the burning of houses, and even murder (often by lynching) became commonplace. After considerable delay, Congress responded with three Enforcement Acts (1870-71), which provided for the protection of the freedmen's right to vote and which were particularly aimed at crushing the KKK. In 1875, a Civil Rights Act was passed, guaranteeing blacks public accommodations equal to those of whites and also providing for black participation on juries. The act lacked any provision for enforcement, however, and was subsequently limited in scope by a Supreme Court decision (1883) ruling parts of it unconstitutional.

Radical Reconstruction was already running out of steam by the time of the presidential election of 1876. Republican Rutherford B. Hayes was pitted against Democrat Samuel J. Tilden in an election that accorded Tilden a majority of the popular vote, but that gave Hayes one more electoral vote than he. Nineteen electoral votes from Southern states were in dispute, and the election went first to a congressional commission and was ultimately decided by a compromise: The Democrats would certify the election of Hayes in return for an end to Reconstruction policies in the South.

The compromise came at the expense of Southern blacks, who fell victim to so-called Jim Crow laws enacted throughout the South—legislation intended to "keep blacks in their place" through segregation and political as well as economic disenfranchisement. Southern racism became institutionalized as white, conservative Democrats universally assumed power throughout the region, creating the so-called Solid South.

CHAPTER 11

INDUSTRIALIZATION, POPULISM, AND PROGRESSIVISM (1877-1916)

THAT THE CIVIL WAR DID NOT, AFTER ALL, DESTROY THE UNITED STATES IS A TESTAMENT to the idea and principle of national union; nevertheless, the nation that emerged after the war was not the nation that had gone into the war. Immigration increased from a steady flow to an overwhelming tide.

The Era of the Immigrant

By definition, the United States was and is a nation of immigrants; yet, during the antebellum (pre-Civil War) years, the pace of immigration, while hardly slow, was steady, and there was little difficulty assimilating immigrants and immigrant culture into what may be called the American mainstream. Not only were immigrant numbers relatively manageable in the prewar years, most immigrants settled on farms, many of them in the newly opened lands of the western prairies and plains. They were not forcibly and immediately thrust into mass American culture.

After the Civil War, the pace of American urbanization rapidly increased, as did the rate of immigration. Many American cities developed a range of ethnic neighborhoods, where immigrants from similar backgrounds tended to live. Such "immigrant ghettos" eased the transition into the American "mainstream," and the communities often developed rich ethnic heritages in the form of newspapers, popular theater, churches, food establishments, and so on. However, the communities also had negative aspects, including isolation from broader economic and social opportunities, which created poverty, slums, and crime.

The relative isolation of the ethnic ghetto was double-edged. On the one hand, it helped ease the transition to life in a new country, but it also invited the prejudice and discrimination of "native-born" Americans. Among this more established group was a range of attitudes. Some believed that the immigrants should be rapidly assimilated and acculturated: they should learn English and acquire an understanding of the American government, legal system, and other institutions. At the other extreme were what historians call "nativists." These individuals (who frequently banded together into such groups as the Protective Association and the Immigrant Restriction League) sought to restrict immigration and to institutionalize various forms of anti-immigrant discrimination.

Urbanization and Industrialization

The late nineteenth-century immigrant theme in American history is closely tied to the rise of the cities and industrialization. In significant measure, it was the North's industrial prowess that won the Civil War, a war fought with mass-produced weapons and supplies and a war that relied heavily on railroad transportation. Following the war, industry, especially in and around the Northern cities, continued to develop rapidly. Between 1860 and 1910, the population of U.S. cities increased by a factor of 7, and, by 1920, the United States was clearly an urban nation, with most Americans living in towns or cities of populations greater than 2,500. Immigrants from southern and eastern Europe contributed mightily to the growth of urban America; increasingly, too, African Americans were leaving the rural South for the urban industrial North.

The rise of the cities brought great innovations in architecture, including the development of the skyscraper, and transportation, including better roads and streets, light interurban railroads, and intraurban transportation in the form of electric trolley lines. The cities also brought about (mostly negative) innovations in politics, in the form of the *political machine*, presided over by one or more *political bosses*. While the machines and their bosses were generally corrupt, they gained and maintained power by serving the needs of the immigrant communities spurned by mainstream politicians. For example, the machine might provide food and support for immigrant widows and orphans, might provide jobs for the unemployed (typically within the city government or infrastructure), and might provide legal representation or cut through bureaucratic red tape. While most

U.S. cities developed political machines, the most notorious, powerful, and corrupt was New York City's Tammany Hall, run by William Marcy "Boss" Tweed. Because machines controlled significant blocs of urban votes, they soon became a force to be reckoned with in state and even national politics.

Robber Barons and Unionism

The tyranny of machine politics was one aspect of the dark side of post-Civil War democracy. The rise of the "robber barons"—powerful, ruthless, and often unscrupulous financiers and industrialists—was another. Through a combination of daring and skill and by exploiting the corruption of the Grant administration, the likes of Jay Gould, Cornelius Vanderbilt, J. P. Morgan, Andrew Carnegie, John D. Rockefeller, and others built huge fortunes on railroads, oil, steel, and other industries. Competition was crushed through the creation of "trusts," so that, within a given industry, one man might control not only the manufacture, distribution, and sale of a product, but the creation of the raw materials as well. Once such total "vertical integration" had been achieved, the "robber baron" could dictate prices and even gouge the public.

General consumers protested high prices and the absence of competition, but it was two classes that suffered most at the hands of the robber barons:

- Farmers were at the mercy of railroad freight rates and often felt themselves strangled by the costs of getting their produce to market.
- Miners and industrial workers found themselves at the mercy of the big companies for which they worked. Often, they were compelled to live in company-owned houses, to buy their goods in company stores, and to live in company towns—typically on marginal wages and in marginal living conditions.

In response to the abuses of the robber barons and big business generally, many farmers banded together in associations known as granges, which grew into a national grange movement, a coalition that fought monopolistic grain transport practices, but that also took on aspects of a political party and even of a secret society. In the industrial sector, the post-Civil War period saw the rise of trade unions.

Among the first great national unions was the Knights of Labor, formed in 1878 and successful in strikes against railroads. In 1886, however, a general strike called in Chicago resulted in the Haymarket Riot, in which several people, including police officers, were killed. The Knights were identified with anarchists, and the organization dissolved. Nevertheless, the late nineteenth century saw more major strikes against monopolistic, worker-exploitive companies such as Carnegie Steel (1892) and the Pullman Palace Car Company (1894). In 1886, the American Federation of Labor was formed and endured as a powerful national union. More radical was the Industrial Workers of the World (IWW), founded in 1905 by Eugene V. Debs and others to address the needs of unskilled labor (the membership of other unions consisted of skilled workers). The "Wobblies," as IWW members were called, vowed permanent class warfare against exploitive employers and spoke of replacing capitalism with a new "industrial democracy." The federal government treated the IWW as a subversive organization.

The Rise of American Popular Culture

The effect of the growth of urban America and great industries was hardly all negative, of course. In general, the standard of living rose, as did the quality of life: health improved and leisure time increased. Americans became a nation of consumers, not only of new manufactured goods (many of them luxury items), but of popular entertainment and sports. The late nineteenth century saw the development of such mass entertainment media as vaudeville, popular theater (including musical comedy, circuses, Wild West shows), and, by the end of the century, phonograph recordings and the movies. The popular press also developed, with "dime novel" fiction making a big hit and sensational journalism drawing a huge following. (During this period, the rival New York papers of Joseph Pulitzer and William Randolph Hearst presided over the creation of the most extreme form of newspaper sensationalism, known as "yellow journalism.") Sport developed from a collection of informal neighborhood or community activities into professional exhibitions. By the early twentieth century, professional baseball became an important and highly profitable business. Later in the century, many other professional sports developed extensive mass appeal as well.

More Westward Expansion

Even as urban America grew, the settlement of the West expanded. During the Civil War, Congress had passed the Homestead Act (1862), authorizing any citizen—or immigrant who intended to become a citizen—to select any surveyed but unclaimed parcel of western public land up to 160 acres, settle it, improve it, and live on it. After five years of occupancy, the homesteader received title to the parcel. Pursuant to the Homestead Act and subsequent legislation, hundreds of thousands of settlers braved the hardships of prairie life to settle the trans-Mississippi West.

While the Homesteaders were farmers, the West also offered vast ranges on which cattle could be raised. When Texans had gone off to fight the Civil War, many of them let their livestock fend for themselves, so that, by the end of the war, millions of head of cattle were ranging free across the state. For many ex-Confederate soldiers, rounding up these Texas strays, branding them, then driving them to market up north became an economic salvation. Almost immediately after the war, the trail drive cattle industry began, and with it came the rise of the cowboy. Although cowboys were typically the poorest of the poor, many without family, friends, or property, and despite the fact their work was dirty, dangerous, lonely, poorly paid, and tedious, these figures become a cherished, powerfully romantic part of American history and popular mythology.

The Indian Wars

Violent conflict between Euro-American settlers and Native Americans began as early as 1493, when Columbus's men clashed with the inhabitants of Hispaniola in the Caribbean. Over the years, despite repeated efforts to "remove" the Indians westward, far from the Euro-American frontier, settlement repeatedly overtook removal, leading to more clashes. Indian warfare played a part in the trans-Mississippi theater of the Civil War, but the so-called Indian Wars began in earnest after the Civil War, in 1866, when the Teton Sioux, the Northern Cheyenne, and the Northern Arapaho attempted to close the Bozeman Trail in an effort to shut off the ceaseless flood of settlers onto their lands. For about the next quarter century, the U.S. Army, various state and local militia groups, and private citizens battled Indians across the whole of the West. In formal terms, the Indian Wars consisted of fourteen military campaigns, culminating in the Battle (or massacre) of Wounded Knee, December 29, 1890.

Contrary to the claims of some Indian as well as white activists in the twentieth century, it was never the federal government's official policy to practice genocide against the native peoples of the West. The official policy was "concentration," the gathering together and installation of Indians on government-regulated and subsidized "reservations." Many of the tragedies that befell United States Indian policy resulted from the almost universal corruption among the agents entrusted with the administration of the reservations. On many reservations, starvation and abuse were the rule. Instead of preventing war, therefore, the reservation system frequently provoked further—and typically desperate—violence. Despite the great skill and daring of Plains Indian warriors, the Native American population in the West was soon vastly outnumbered. Against approximately ten-to-one odds, no warriors, regardless of skill and commitment, could long prevail.

Populism

Collectively, the political protest movements of the Grange and of the labor unions are often called Populism. In 1891, the Farmers' Alliances, a movement that ultimately supplanted the Grange, motivated the formation of the Populist Party. The Populists advocated free and unlimited coinage of silver (to increase the money supply), government ownership of railroads and telegraph and telephone lines, a graduated income tax (a tax level based on individual income), and more direct public participation in legislation by means of initiative, referendum, and recall.

While the Populist Party disappeared after the elections of 1908, Populist sentiment remained strong among the laboring and agrarian classes and persisted as a significant influence in "mainstream" American politics.

Progressivism

By the turn of the century, Progressivism, a movement allied to Populism, but more sophisticated and associated with the liberal elite more than with laborers and farmers, came into being. Indeed, the turn of the century is often called the Progressive Era.

The Progressives were avid reformers, who advocated more government control of business, yet who were less radical than the Populists in that they did not want to abandon the basic principles of capitalism and free enterprise. The Progressives:

- Lobbied for government regulation of business and industry to end abusive monopolies
- Championed *progress*—the continued growth and technological and social advancement of the United States
- Worked to clean up corruption at all levels of government

◆ Advocated government intervention in a variety of community affairs

In an atmosphere of reform, various religious organizations (including the Salvation Army) promoted social responsibility and good works. The tradition of the crusading journalist was born in this period. "Muckrakers" was what Theodore Roosevelt (a leading Progressive politician) called such writers as Ida M. Tarbell (who exposed the corruption of Standard Oil Company), Lincoln Steffens (author of *The Shame of the Cities*), Jacob Riis (photographer and author of *How the Other Half Lives*, documenting tenement life in New York), and Upton Sinclair (whose novel *The Jungle* exposed the abuses of the meat-packing industry of Chicago). (The word *muckraker* came from a character in John Bunyan's classic moral allegory *Pilgrim's Progress*, who rejected a crown for a muckrake, with which he gathered and cleared away dung.) It was during this period that the profession of social work was born, pioneered by such reformers as Jane Addams, whose Hull House in the slums of Chicago was dedicated to improving the lot of poor immigrants. Finally, Progressivism saw the rise of women's clubs and associations, which became the core of a renewed feminist movement and a drive for women's suffrage.

Progressivism has sometimes been criticized for actually reinforcing the social status quo—effectively keeping the "lower classes" in their place by placating them. Nevertheless, the Progressive tradition measurably improved the lot of women, of laborers, of urban slum dwellers, and especially children—protecting them from dangerous and exploitive working conditions. Because of Progressivism, the government took a much more active role in social legislation and in regulating industries that produce food and drugs. Progressivism shaped the "Square Deal" policies of President Theodore Roosevelt and, later, would influence the Depression-era reforms of Franklin D. Roosevelt's "New Deal."

American Imperialism: The Spanish-American War

In February 1896, General Valeriano Weyler arrived in Havana as the new governor of the restless Spanish colony of Cuba. He set about crushing rebellion, rounding up Cubans and incarcerating them in "reconcentration" camps. Many Americans sympathized with the rebel cause, and the reporters of newspaper rivals Pulitzer and Hearst vied with one another to report atrocity stories. At last, under intense public pressure, President William McKinley dispatched the battleship *Maine* to Havana harbor "to protect American citizens" in Cuba. On February 15, 1898, the ship suddenly blew up, killing 266 sailors. Spanish authorities were blamed (although it is now believed that the explosion was internal and accidental), and, amid cries of "Remember the *Maine*!", the United States tottered on the verge of declaring war on Spain in order to annex Cuba. The Spanish beat the Americans to the punch, declaring war on the United States on April 24.

In what Secretary of State John Hay called "that splendid little war," U.S. land and naval forces forced Spain to withdraw from Cuba by August. The United States did not annex Cuba—though its government remained closely tied to the U.S.—but did acquire from Spain Puerto Rico and Guam, and also purchased the Philippine Islands. While support for the war and the subsequent territorial expansion was overwhelming in the United States, there was a significant minority who criticized the entire affair as an example of a strong-arm imperialism that ran contrary to American ideals.

The Era of Teddy Roosevelt

The son of a privileged New York City family, Theodore "Teddy" Roosevelt entered politics at age twenty-three as a state assemblyman, quickly earning a reputation as an opponent of corruption. After suffering some political defeats, he spent two years ranching in the Dakota Territory, then became a reforming member of the U.S. Civil Service Commission (1889-95) and president of the New York City Board of Police Commissioners (1895-97). As assistant secretary of the Navy under President McKinley, he had urged war with Spain, then resigned the post to organize the 1st Volunteer Cavalry, better known as the Rough Riders. Roosevelt led a famous charge up San Juan Hill in the war's most critical land battle. Returning to the U.S. a tremendously popular figure, Roosevelt was tapped by Republican party boss Thomas C. Platt to run for governor of New York. When, as governor, Roosevelt began removing corrupt politicians from office and enacted other reforms over Platt's opposition, the party "kicked Roosevelt upstairs," nominating him for vice president on the McKinley ticket.

In 1901, McKinley was assassinated by an anarchist named Leon Czolgosz, and Roosevelt assumed office. He initiated a raft of Progressive reforms:

- He moved against industrial and commercial "trusts."
- He created the Department of Commerce and Labor to regulate business

and enforce economic regulations.

- He created the Bureau of Corporations, which later became the Federal Trade Commission
- He strengthened federal regulation of interstate commerce.

In running for election in his own right in 1904, he promised Americans a "Square Deal," a pledge to use his office to ensure public safety, prosperity, and happiness. His second term was marked by passage of more regulatory legislation, including the Meat Inspection Act (1906) and the Pure Food and Drug Act (1906).

Roosevelt's Imperialism

At the end of the nineteenth and beginning of the twentieth century, American foreign policy was influenced by what some have called the New Manifest Destiny. In the mid nineteenth century, the nation's "manifest destiny" was seen as conquest of the continent. At century's end, with the United States indeed extending from sea to sea, the *new* manifest destiny looked beyond U.S. borders. The Spanish-American War was one dramatic result of this imperialist impulse, as was a desire to unite the hemisphere through a series of Pan American Conferences and various cooperative agreements. In the Pacific, American interests in Hawaii moved the islands toward U.S. annexation, which occurred in 1898. England, Germany, and the United States disputed possession of Samoa, and in 1899 the islands were divided between the U.S. and Germany.

Under Roosevelt, the Monroe Doctrine was reinterpreted as a sanction for a good deal of U.S. intervention in Latin American affairs, including control over the soon-to-be-built Panama Canal (opened in 1914).

During the administrations of McKinley and Roosevelt, the United States also took a strong interventionist stance in Asia. An attempt was made to put down a Filipino bid for independence, and in 1902 the islands were made an unincorporated U.S. territory. It was not until 1916 that the Jones Act promised Filipino independence when a stable government was established (which took place in 1946). In 1900, United States troops acted in concert with European and Japanese forces to put down the so-called Boxer Rebellion in China, an attack on European civilians in Peking (Beijing). Yet in 1905, Theodore Roosevelt played peacemaker in negotiating the Treaty of Portsmouth, ending the Russo-Japanese War. (Roosevelt was subsequently awarded the Nobel Peace Prize.)

The Conservation Movement

Among Roosevelt's greatest legacies to the nation was a strong conservation policy, aimed at stemming the destruction of the American wilderness. Much important conservation legislation was launched, and the foundation of the modern conservation movement laid.

Social Retreat Under Taft

Roosevelt did not seek reelection in 1908 and handpicked his secretary of War, William Howard Taft, to stand as the Republican candidate. To the dismay of Roosevelt and other progressives, however, the Taft administration proved quite conservative and generally retreated from TR's social activism. In response, Roosevelt joined the Progressive (or Bull Moose) Party and ran unsuccessfully against Republican Taft and Democrat Woodrow Wilson.

"New Freedom" Under Wilson

Although he was a Democrat, Woodrow Wilson, former professor and president of Princeton University and governor of New Jersey, shared elements of Roosevelt's Progressivism, but took it in a more liberal direction, which his supporters called the "New Freedom." Wilson advocated:

- Not merely the breakup of trusts, but the elimination of all monopolies and the strict regulation of big business
- A graduated income tax
- Government regulatory roles in social and economic affairs
- A Federal Reserve system to regulate and stabilize currency
- Strong child labor reform
- Legislation to fund vocational education

Wilson's Progressive activism is evidenced by his nomination of the brilliant Louis Brandeis to the Supreme Court (the first Jew so nominated) and by the passage and ratification of four constitutional amendments, including income tax (16th Amendment, 1913), the direct election of Senators (17th Amendment, 1913), prohibition of alcoholic beverages (18th Amendment, 1919—enacted over Wilson's objection), and the right of women to vote (19th Amendment, 1920).

CHAPTER 12

FROM WORLD WAR I TO WORLD WAR II (1916-1945)

UNDER ROOSEVELT'S SUCCESSOR, WILLIAM HOWARD TAFT, "DOLLAR DIPLOMACY" promoted U.S. business interests overseas, with the government authorizing diplomatic and military protection for American enterprises throughout Latin America. President Wilson pursued much the same course, establishing a military government in the Dominican Republic and in Haiti. Mexico deeply resented Wilson's high-handed Latin American policies, bringing U.S.-Mexican relations to their lowest point since the U.S.-Mexican War of the previous century. Yet despite certain imperialist gestures under McKinley, Roosevelt, Taft, and Wilson, most Americans were essentially isolationists. They put stock in the advice George Washington gave in his Farewell Address: "beware of foreign entanglements."

Isolationism and Its End

U.S. isolationism was in large part a function of geography. The nation was separated from Europe and Asia by vast oceans, and it was far more powerful than its northern neighbor, Canada, and its southern neighbor, Mexico. Isolationism was put to its severest test by the outbreak of World War I, which began on June 28, 1914, with the assassination of the Austrian Archduke Francis Ferdinand and his wife Sophie, while they were visiting Serbia. The incident activated a complex network of interlocking alliances (including a number of secret treaties) among European nations, bringing Germany and the Austro-Hungarian Empire (called the Central Powers) into war against France, Great Britain, and Russia (called the Allies). (Later in 1914, the Ottoman Empire joined the Central Powers, and Italy joined the Allies.)

President Wilson reacted to the "European War" by seeking to maintain strict U.S. neutrality. This was especially difficult because:

- Most Americans (Wilson included) felt a strong cultural bond with the English.
- Trade with France and Britain in war materiel was highly profitable.
- Stories of German war atrocities (especially the wholesale killing of civilians in Belgium) turned public opinion against Germany.
- Germany repeatedly violated U.S. rights as a neutral.

Violations of neutrality included, most importantly, the 1915 sinking of the British liner *Lusitania* by a German U-boat; 128 Americans were among those lost at sea. When a French vessel, the *Sussex*, was attacked in 1916, resulting in the injury of several Americans, Wilson extracted from the Germans a pledge not to torpedo civilian passenger vessels without first giving warning. This pledge was broken by January of the next year. By the beginning of 1916, a preparedness movement swept the United States, and Wilson and Congress authorized rearmament later in the year. In 1916, Wilson won reelection on the slogan "He kept us out of war," but in January 1917, British authorities intercepted a telegram from Alfred Zimmerman, German foreign secretary, to the German ambassador to Mexico. It authorized the ambassador to urge Mexico to support Germany if the United States declared war on Germany; for this support, Germany pledged to secure the return to Mexico of Texas, Arizona, and New Mexico. The British turned over the "Zimmerman note" to Wilson, who was outraged. The next month, after Germany announced the resumption of unrestricted submarine warfare—open season on all ships—Wilson resolved to ask Congress for a declaration of war. This was secured on April 6, 1917.

America Enters the "Great War"

The United States came into the war at a low point for the Allies. At the time of the war declaration, the U.S. Army numbered about 200,000 men. During the war, that number swelled to four million, including 2.8 million draftees. The

American Expeditionary Force (A.E.F.) was commanded by General John J. ("Black Jack") Pershing, while naval forces were under the direction of Admiral William S. Sims. Pershing arrived in Paris on June 14, 1917, but no U.S. units were committed to the front until October 21, and it was not until the spring and summer of 1918 that massive numbers of American soldiers entered into the fighting as the Germans launched a series of desperate offensives along the Western Front from March through June. From June 6 through July 1, U.S. forces recaptured Vaux, Bouresches, and Belleau Wood from the Germans, and Americans held the town of Cantigny against a massive German offensive threatening Paris during June 9-15. Between July 18 and August 6, a force of 85,000 American troops crushed the last major German offensive at the Second Battle of the Marne, the turning point of the war.

The U.S. Army fought in the Somme, Oise-Aisne, and Ypres-Lys offensives of August 1918, and at St. Mihiel Salient (September 12-16). From the end of September through the day of the Armistice, all available U.S. forces were concentrated along a sector between the Meuse River and the Argonne Forest. The success of this offensive brought about the capitulation of Germany, and an armistice was declared on November 11, 1918. The definitive Treaty of Versailles was concluded on June 28, 1919.

U.S. forces suffered deaths numbering 112,432 (about half not from bullets, but from an influenza-pneumonia pandemic), and wounded numbering 230,074. In money, the war had cost Americans more than $20 billion.

Fourteen Points and a League of Nations

President Wilson was determined that the "Great War," so costly, would be the last world war Americans would fight. Accordingly, as early as January 8, 1918, he announced to Congress "Fourteen Points . . . as the only possible program" for peace, as far as the interests of the United States were concerned. The Fourteen Points included provisions for open covenants of peace (no secret treaties and alliances such as those that led to war in the first place), freedom of the seas, withdrawal of invading armies and restoration of occupied territories, adjustment of borders, and, most significantly, the establishment of a general association of the world's nations for the purpose of ensuring the political independence and territorial integrity of states great and small.

Wilson played a key role in the peace conference at the end of the war, and his Fourteen Points were incorporated into the Treaty of Versailles, as was a provision and charter for the creation of a League of Nations—an international body dedicated to the peaceful resolution of disputes. But American isolationist factions reemerged after the war, and, despite Wilson's vigorous campaigning, the Senate refused to ratify the Treaty of Versailles or permit the United States to join the League of Nations. Wilson threw himself body and soul into the battle on behalf of the treaty and the League, touring the nation in an effort to raise popular support. The struggle broke his health, he suffered a debilitating stroke, and served out the balance of his term as an invalid.

The Lost Generation and the Roaring Twenties

Postwar America was not so much feeling flushed with victory following the Great War as it was plagued by strikes, beset with the effects of a worldwide influenza pandemic, and gripped by fears of Communism. In 1919, reacting to the "Red Scare," U.S. Attorney General A. Mitchell Palmer appointed a young J. Edgar Hoover (soon to become famous as director of the FBI) to stage a series of raids on suspected Communist sympathizers. Private homes as well as union headquarters and social-political clubs were raided. Some 6,000 so-called radicals were arrested and held without charge, in violation of their civil rights. Almost 300 were deported to the Soviet Union.

In the wake of the Red Scare, two Italian immigrant workers, Nicola Sacco and Bartolomeo Vanzetti, were arrested for murder in South Braintree, Massachusetts. Admitted radicals, they were found guilty in a clearly prejudiced trial, and their subsequent appeal became a national and international cause in which prominent political liberals, writers, humanitarians, and statesmen, including President Wilson, sought their retrial. Despite these protests, the two were executed in 1927.

The nation, reeling from war and postwar political chaos in Europe, was eager to retreat into isolationism and what Harding called a "return to normalcy." Not only was Wilson's League of Nations rejected, but Warren G. Harding, an affable Republican conservative, was elected to the presidency, in large part on an isolationist platform. "The League of Nations," he said in one campaign speech, "isn't for us."

Like Ulysses S. Grant, Harding seemed blissfully unaware of the wholesale corruption that quickly engulfed his administration—most notably the Teapot Dome scandal (1924), in which Secretary of the Interior Albert B. Fall was found guilty of having secretly leased U.S. oil lands (reserved exclusively for the U.S.

Navy) to oil magnate Harry F. Sinclair in return for a "loan."

After Harding's death in office in 1923 (apparently from accidental food poisoning, though some have suspected homicide), Calvin Coolidge completed his term and was elected to a term in his own right. The Coolidge administration was not marked by scandal, but was notable for its inaction: no significant legislation was proposed, and no foreign policy initiatives introduced. The taciturn Coolidge was popularly called "Silent Cal."

The wholesale corruption of the Harding years and the laissez-faire approach of the Coolidge presidency were very much in keeping with the attitude of the times—the "Roaring Twenties"—emotionally burned out from war and seeking wild escapism in jazz, loose morals, and booze. Made illegal by passage of the 18th Amendment (1919), liquor flowed plentifully.

The 1920s saw the emergence of important American artists and writers (most notably the novelists Ernest Hemingway, F. Scott Fitzgerald, William Faulkner, the African American poet Langston Hughes, the playwright Eugene O'Neill), whom the writer Gertrude Stein collectively characterized as the "Lost Generation." This description seemed apt applied to all who came of age in what seemed a spiritually desolate era, its ideals having been shattered by war and crisis.

Prohibition and the Emergence of Organized Crime

Prohibition—the "noble experiment"—became the law of the land in 1919 with passage of the 18th Amendment. Immediately, a vigorous trade in "bootleg" alcohol developed and, around it, a new class of criminal. In the past, American criminals were perceived either as lone misfits—social aberrations—or as romantic Robin Hoods (Billy the Kid or Jesse James). But trade in illegal liquor was a big business, and the hoodlums of the 1920s began to model themselves on legitimate businessmen. "Organized crime" came into being and, with it, such figures as Deanie O'Bannion, Al Capone, Dutch Schultz, and many others. Organized crime reached into every level of government. Cities like Chicago and New York were riddled with corruption, as gangsters bribed police officials, judges, city councilmen, and even mayors. The federal government stepped in to combat crime, and, under J. Edgar Hoover, the Federal Bureau of Investigation grew into a powerful crime-fighting organization, the closest thing the nation has ever had to a federal police force.

New Prosperity

After a brief recession in 1920-21, the nation entered a period of unprecedented prosperity, in which the gross national product increased by some 40 percent and unemployment dropped as low as 3 percent. Many investment fortunes were quickly made. Yet, in some basic ways, the prosperity was illusory:

- Farmers, hit hard by the postwar recession, never shared in the general prosperity of the decade.
- Ordinary laborers made few real gains.
- Much of the investment wealth was built on shaky foundations.

Manufacturers in the 1920s turned out a vast array of consumer goods, ultimately producing far more than consumers could buy. This created an instability that set up the national economy for a fall. In the meantime, the general public leaped into the stock market, often making careless investments and borrowing heavily to do so. Stocks were soon overvalued.

The Rise of Women

Despite the conservatism suggested by Prohibition and embodied in the policies of Harding and Coolidge, the 1920s took a fresh look at morality, social mores, racial equality, and the role of women. Although white intellectuals took a keen interest in the literary and artistic activity of the writers and artists of the Harlem Renaissance (a cultural movement centered in New York City's major black district), the majority of African Americans still endured prejudice and discrimination. Black nationalism became important during this period, with the coming into prominence of Marcus Garvey, who led a return to Africa movement and, in the process, sought to imbue black Americans with a sense of pride, not shame, in their heritage and struggle.

The most dramatic social change during the decade, however, was in the social role and political status of women. In 1920, ratification of the 19th Amendment finally gave women the vote, and women were also emerging as the professional and social equals of men in other aspects of life. The popular image of the "liberated woman" of the 1920s was the *flapper*, a woman who made her own way economically and who not only engaged in such "unfeminine" activities as smoking and drinking, but took an aggressive role in sexual relationships.

The Crash of 1929 and the Great Depression

The economic instabilities and excesses of the 1920s led to a stock market crash in October 1929, which heralded a severe decade-long depression. Contributing causes of the Great Depression included:

- A poorly diversified industrial base
- Too many goods versus low consumer demand (most workers and farmers were too poor to afford many consumer luxuries)
- Poorly regulated banks
- Decreased demand for U.S. exports
- Unpaid European war debts to U.S.

The Depression brought high unemployment combined with failures of investments and banks. Many families found themselves not only without employment, but with their life savings suddenly wiped out. Investors, especially those who had purchased stock "on margin" (on credit), also found themselves financially ruined.

Herbert Hoover, who had entered the White House in 1929, was slow to respond with federal action to alleviate the economic crisis. He encouraged many voluntary steps from private industry and the states, and only belatedly introduced a number of federal programs—but stopped short of granting federal aid directly to individuals because he was concerned that doing so would compromise individual initiative.

With Communism and radicalism in the air during this period of crisis, there was real fear that the economic crisis would bring about the collapse of American democracy. In June 1922, 20,000 unemployed World War I veterans marched on Washington, D.C., to petition Congress to release a "bonus" promised for 1945. This "Bonus Army" was violently dispersed by troops under the command of General Douglas MacArthur. The action made Hoover highly unpopular, and he lost his bid for reelection in 1932 to the charismatic, upbeat, ever-optimistic Franklin D. Roosevelt.

FDR and the New Deal

Where Hoover had been slow to act to introduce federal assistance, FDR moved swiftly through the first "Hundred Days" of his administration, ushering through Congress a series of important laws that formed the foundation of what Roosevelt called the New Deal. In addition, FDR used the radio to conduct "fireside chats" with the American people in a highly successful effort to reassure them that the crisis would be dealt with and their welfare ultimately restored.

The New Deal included such measures as:

- A four-day bank holiday to stave off panic and failures
- Emergency banking legislation to reopen failed banks
- Longer-term Federal Reserve legislation, to strengthen the Federal Reserve and improve regulation of banks
- Creation of the Federal Deposit Insurance Corporation, to insure deposits
- Reform of the securities industry
- Measures to purchase silver and gold in order to stabilize currency
- Reform of income tax, increasing the burden on the wealthy
- Programs to provide emergency relief for individuals
- Federal mortgage programs
- Civil Works Administration, Public Works Administration, and Civilian Conservation Corps: federal programs to create public-works jobs
- Social Security Act
- Agricultural Adjustment Act, Farm Credit Act, Tennessee Valley Authority, and Rural Electrification Administration: programs to aid the long-suffering farmer
- National Industrial Recovery Act: this created the National Recovery Administration (NRA) to regulate industry and labor relations

Although the New Deal did not end the Depression, its many programs did bring relief to individuals and industries, and political turmoil, civil unrest, and revolution were averted. For better or worse, the New Deal permanently cast the federal government in a socially active role.

The U.S. Enters World War II

Most of the rest of the world was not as fortunate as the United States. The Great Depression was worldwide, and throughout Europe economic hardship created government instability. Strongman dictators Benito Mussolini (Italy) and Adolf Hitler (Germany) rose to power. In Japan, a military dictatorship took hold.

Partly in the name of fighting the onslaught of Communism, Italy, Germany, and Japan banded together (becoming "the Axis" powers). Throughout the 1930s, the Axis rearmed (in defiance of the Treaty of Versailles) and positioned itself for world domination:

- Germany and Italy aided fascist leader Francisco Franco in the Spanish Civil War, ensuring there the establishment of a government friendly to Mussolini and Hitler.
- Italy bombed and invaded Ethiopia. Ethiopian emperor Haile Selassie appealed to the League of Nations for aid, but the League proved powerless.
- Japan invaded and annexed Manchuria.
- Germany annexed Austria and the Sudetenland of Czechoslovakia.

Despite strong isolationist sentiment in the United States, FDR authorized preparations for possible war. Neutrality Acts were passed, then, late in the 1930s, revised to allow shipment of arms to England and France. In 1940, the Selective Service and Training Act commenced the first peacetime military draft in American history. Later in the year and early in 1941, programs of direct arms shipments to Britain and the Soviet Union were authorized. Finally, the Japanese air attack on the U.S. naval base at Pearl Harbor, Hawaii (December 7, 1941), brought the United States into World War II (against Japan on December 8 and Germany and Italy on December 11). With incredible speed, American industry bent to the war effort—and the effects of the Depression were quickly dispelled.

World War II: The Pacific Theater

Although, for the United States, the war started in the Pacific, military and political leaders saw the European front as their first priority. These official war aims did not match the mood of the American public, which regarded the war chiefly as a means of taking revenge on Japan. Indeed, Japan had not only surprised the United States at Pearl Harbor, but had also attacked U.S. forces in the Philippines, quickly overrunning them. The first counterstrike against Japan was more a desperate gesture than a militarily strategic move. On April 18, 1942, Lieutenant Colonel James Doolittle led sixteen Army Air Force B-25s in a spectacular low-level surprise attack against Tokyo. Although only minor damage was done and all aircraft lost, most of the crews were saved, the morale of the

United States was greatly lifted, and the stunned Japanese were compelled to keep some valuable air assets at home. Of far greater strategic consequence was the naval Battle of Midway (June 4-6, 1942), Japan's first defeat in the Pacific. After Midway, U.S. ground, air, and naval forces steadily began to reclaim control of the Pacific.

World War II: European Theater

U.S. forces cooperated with the British in an invasion of North Africa in 1943, from which an invasion of Sicily and mainland Italy was launched. Simultaneously, the United States and its British allies commenced a terribly costly strategic bombing campaign against German and Italian forces in Europe. It was not until June 6, 1944 (D-Day), that the main Allied invasion of what Hitler called Fortress Europe was commenced with the greatest amphibious assault in history. Under overall command of General Dwight D. Eisenhower, the troops struck at Normandy. Once the hard-fought beachheads were secured, the U.S. Third Army, under General George S. Patton, led a lightning advance across Europe, steadily forcing the Germans back toward their homeland. A desperate counteroffensive, begun on December 15, 1944 (the Battle of the Bulge), proved the war's most costly engagement for the Americans, but, after retreating and regrouping, U.S. troops ultimately routed the Nazis. The European war ended on May 8, 1945 (V-E Day), with the surrender of Germany.

The Atomic Age Is Born

While U.S. forces made steady gains in the Pacific, the Japanese showed themselves to be an incredibly stubborn enemy, literally fighting to the death and even using suicide tactics: kamikaze pilots willingly crashed their explosive-laden aircraft into American ships. Beginning early in the war, a group of American scientists under the scientific direction of J. Robert Oppenheimer and the military direction of General Leslie R. Groves had been working on the "Manhattan Project"—the creation of an atomic bomb. In the summer of 1945, the weapon was ready for use, and President Harry S Truman (who had assumed office after FDR, in his fourth term, succumbed to a cerebral hemorrhage on April 12, 1945) made the fateful decision to employ it against Japan. Two bombs were dropped, one on Hiroshima (August 6) and another on Nagasaki (August 9), whereupon the Japanese surrendered, and World War II came to an end.

The Homefront

The war brought profound social changes to America. The labor demands of war industries accelerated the migration of African Americans from the rural South to the cities of the industrial North. Racial tension often escalated, and U.S. cities saw a series of race riots during the war years. However, the cause of civil rights was greatly advanced during this period, and while the armed forces (except for the navy) were still officially segregated during World War II, blacks and whites did serve together. For many of both races, it was the first prolonged, cooperative contact with one another.

Women went to work in war industries in unprecedented numbers. For many, this was their first taste of a world beyond home, husband, and children.

Japanese-Americans became the target of racial prejudice and, by federal order, those living on the West Coast were "relocated"—removed from their homes and placed in internment camps for the duration of the war.

Despite racial turmoil and social upheaval, the U.S. homefront was remarkably unified during World War II, which many Americans fondly recall as a time of intense emotional commitment, solidarity, and patriotic purpose in a contest, quite simply, of good against evil.

CHAPTER 13

POSTWAR AMERICA (1945-1968)

REJOICING AT THE END OF WORLD WAR II WAS BRIEF. THE SOVIET UNION, AN ALLY against the Nazis, was once again a political and ideological enemy in a world divided between the western democracies and the eastern Soviet "bloc." The Hiroshima bomb had ushered in an atomic age, and the world soon seemed poised on the brink of World War III—a nuclear war that might well end civilization. On the hopeful side, however, the United States vigorously supported the new United Nations, an international organization that promised to succeed, as an alternative to war, where the League of Nations had failed.

Social Revolutions

On the domestic front, the postwar years brought new prosperity as consumer demand, pent up by wartime rationing and scarcities, mushroomed. A "baby boom" got under way as returning veterans married and started families. Around American cities, vast suburbs developed to house the growing population. The G.I. Bill of Rights provided to returning veterans inexpensive home loans and education loans, among other benefits. In economic terms, the prospects for postwar America looked bright.

Blacks, women, and other minority groups generally benefitted during the postwar years. With booming consumer demand came greater employment opportunity, and even though racial prejudice continued to be a major problem in American life, the war experience tended to broaden horizons for blacks as well as whites. Employment and social opportunities gradually opened up.

Truman, Korea, and the Cold War

Acting on the advice of diplomat George F. Kennan, President Truman developed a foreign policy whose keynote was "containment": the United States would act to check (to "contain") the expansion of Communism wherever it appeared in the world. This policy was embodied in the so-called Truman Doctrine, which proposed economic and military aid to any nation threatened by Communism. The United States was now a long way from its traditional isolationism.

In large part to preserve from Communist domination as much of Europe as possible, Secretary of State George C. Marshall proposed a mammoth foreign aid program that came to be called the Marshall Plan. In effect, the United States would help finance the rebuilding of Europe. Not only was this a great humanitarian gesture, it was also a politically savvy program in that it cemented allegiances with the West instead of with the Communist East.

Under Truman the Rio Treaty (1947) was signed, forming a mutual defense alliance among the nations of the Western Hemisphere, and the North Atlantic Treaty Organization (NATO) was created (1949), an alliance between the United States and the countries of Western Europe.

The political and ideological lines were drawn, and the world settled into a long "Cold War," with the United States and its allies facing down the Soviet Union and its allies. Germany had been divided after the war into western and eastern zones of occupation, the western zone controlled by the United States, France, and England, and the eastern zone by the Soviet Union. When the Western nations merged their zones into a new nation, West Germany, the Soviets responded by blockading Berlin. That city was squarely within Communist-controlled East Germany, but had been divided into western and eastern zones. In response to the Soviet blockade, Truman ordered a massive round-the-clock airlift of food and other supplies to West Berlin. The airlift went on for almost a year until the USSR's leader, Josef Stalin, lifted the blockade.

The triumph of the Berlin Airlift was quickly tempered by the failure of the "containment policy" in Asia. The Nationalist Chinese forces under U.S. ally Chiang Kai-shek were decisively defeated in 1949 by Communist Chinese forces under Mao Tse-tung. China, the largest nation in the world, was now a Communist giant.

In 1948, Korea was divided into a Communist North and a democratic South. Two years later, North Korean forces invaded the South. Acting in accordance with a U.N. directive, the United States fought an undeclared war (called a "police action") against the North Koreans, pushing them back to the border of China. The general in charge of U.N. forces in Korea, Douglas MacArthur, hero of the Pacific theater of World War II, proposed expanding the war by attacking the Chinese, who were clearly aiding the North Koreans. President Truman, fearing an escalation that might expand into a third world war, blocked MacArthur. When MacArthur publicly criticized Truman's policy, Truman, as commander in chief, removed the general from command. Ultimately, the war was fought to a stalemate and an armistice concluded, establishing a cease-fire line at the 38th parallel, the border separating North from South.

The McCarthy Era

The Truman era was marked by the president's liberal social policies (the "Fair Deal") on the one hand (which included expanded Social Security, welfare, and labor legislation) and reactionary, repressive conservatism on the other. The House Un-American Activities Committee (HUAC) had been established in 1938 under FDR to disclose foreign influences in the prewar United States. In 1947, Republicans used the congressional committee to associate Democratic-controlled government with Communist subversion. HUAC investigated a number of government agencies and industries—most notoriously the Hollywood film industry, charging that it was riddled with Communist sympathizers ("fellow travelers") and that films contained Soviet propaganda. A "blacklist" of individuals identified by the committee as having ties to Communism was created, and many careers were ruined as a result. In 1950, Senator Joseph McCarthy, Republican from Wisconsin, emerged as the leading hunter of Communists in government. His reckless charges and highly publicized Senate hearings violated the civil rights of scores of accused Communists and Communist sympathizers. McCarthy leveled charges of Communist influence, most of them unfounded, against not only the Democratic Truman administration, but the Republican administration of Dwight D. Eisenhower and the U.S. Army. At length, by 1954, McCarthy's extremism was exposed, and, on December 2, 1954, the Senate voted his censure.

As the Sacco and Vanzetti case had become a focal point of protest against the injustice of the Red Scare in the 1920s, so the case of Julius and Ethel Rosenberg, a husband and wife convicted of giving hydrogen bomb secrets to the Soviets, became the focus of similar protests in the 1950s. Many felt that the conviction was unjust and that, in any case, even if well founded, did not warrant execution of a father and mother. Despite appeals and protests, the couple was executed in 1953.

The Civil Rights Movement

Under President Truman, the civil rights movement received a cautious but important endorsement. His President's Committee on Civil Rights advocated stronger civil rights laws (1946), and while none were enacted during the Truman administration, the president did issue an executive order desegregating the U.S. armed forces (1947). This was an important beginning to the integration of American society at large. Truman also appointed a number of African American judges to the federal bench, and he issued an order prohibiting racial discrimination in government employment.

The Eisenhower Years

Although President Truman could have run for reelection (he had served only one term as an elected president), he chose not to. Republican Dwight Eisenhower, the universally admired architect of Allied military victory in Europe, easily defeated Democratic candidate Adlai E. Stevenson. He offered the nation a hands-off presidency and an administration that supported big business. During his two terms as president, the nation experienced economic well-being, and if Eisenhower was hardly a social activist, much important social legislation was passed during this period, which also saw key Supreme Court decisions:

- In 1954, the Supreme Court ruled the racial segregation of public schools unconstitutional (*Brown v. Board of Education of Topeka, Kansas*). This led to the use of federal troops to enforce desegregation in Little Rock, Arkansas (1957).
- Rosa Parks, an African American, was arrested in 1955 for failing to relinquish to a white passenger a seat in the front of a Montgomery, Alabama, city bus. The black community initiated a nationally publicized boycott of

the Montgomery transit system, and the Supreme Court ultimately ruled segregation on public transit systems unconstitutional.

- During the 1950s, Dr. Martin Luther King, Jr., emerged as the nation's most visible civil rights leader.
- Eisenhower did not introduce any bold civil rights initiatives, but he did complete the integration of the armed forces, which had begun under Truman, and he signed into law a tepid Civil Rights Act (1957).

Other important social legislation during the Eisenhower years included:

- Creation of the Department of Health, Education, and Welfare (1953)
- Extension of Social Security
- A Housing Act provided for low-cost public housing
- The Federal Aid Highway Act (1956), which began construction of a vast network of interstate superhighways

In foreign policy matters, Eisenhower continued Truman's program of containment of Communism. His fiercely anti-Communist secretary of state, John Foster Dulles, developed harsh, uncompromising, and even terrifying policies on the use of the nuclear threat. He advocated "massive retaliation," pledging to answer any attack on the United States with an all-out nuclear response. There now developed a nuclear arms race between the United States and the Soviet Union, as each country amassed stockpiles of weapons of mass destruction. As for Eisenhower himself, he expressed the policy of containment in terms of a "domino theory." Citing the situation of Vietnam, a small nation in southeast Asia that few Americans had even heard of, he advocated bolstering the friendly regime of Ngo Dinh Diem against the Communist leader Ho Chi Minh, arguing that, if Vietnam became Communist, the rest of Asia would follow, just as a row of dominoes falls when one is pushed over.

Late in his presidency, Eisenhower faced two crises: the fall of Cuba to the Communist leader Fidel Castro (which ultimately resulted in a severance of diplomatic relations between the United States and the island nation in 1961) and the U-2 spy plane incident, in which an American high-altitude surveillance aircraft was shot down over the Soviet Union. The incident caused the cancellation of an important U.S.-Soviet summit meeting and significantly chilled already precarious U.S.-Soviet relations.

JFK's "New Frontier"

As the mood of the American electorate had shifted from the activism of Truman to the status quo advocacy of Ike Eisenhower, so it was attracted in 1960 to a new political personality and style. John F. Kennedy, young, energetic, and dynamic, seemed to challenge Americans to greatness. His victory over Richard M. Nixon (vice president under Eisenhower) was narrow, and Kennedy's Catholicism figured as an election issue. (He was the first Catholic president.) Many historians believe that, for the first time in history, television played a major role in a presidential election; four nationally televised debates revealed a confident Kennedy and an apparently stressed Nixon. This image may have tipped the balance in Kennedy's favor.

Where Teddy Roosevelt had presented a national program called the Square Deal, Wilson the New Freedom, FDR the New Deal, and Truman the Fair Deal, Kennedy proposed the New Frontier, a package of initiatives that included:

- Aid to public schools
- Expanded welfare legislation
- Expanded civil rights legislation
- Environmental legislation
- Medicare legislation

Yet most of Kennedy's social program met with strong opposition in Congress and would not be enacted until the administration of Lyndon B. Johnson.

Major programs that are key legacies of the Kennedy years include:

- **The Peace Corps:** a bold international goodwill program that recruited men and women of all ages (though mostly young) to give hands-on assistance to communities in developing nations.
- **The accelerated development of the American space program:** In 1961, JFK set a goal of putting a man on the moon before the end of the decade—in large measure to defeat the USSR in the ongoing "space race," but also to demonstrate American excellence, and simply to fulfill an important dimension of the human spirit of adventure and curiosity. The sixties were an intense period of manned space exploration, and the lunar landing was achieved in 1969.

Cuban Missile Crisis

If Kennedy had a difficult time with his domestic programs, his administration faced even more serious challenges in the area of foreign policy. He sought to strengthen positive relations with Latin America through the Alliance for Progress (1961), essentially a cooperative aid program with Latin American nations aimed at deterring the spread of Communism in the hemisphere. In this sense, it may be seen as an extension of the Monroe Doctrine.

Yet it was precisely in a Latin American nation that the gravest crises of the JFK administration developed. On April 17, 1961, a CIA-trained force of anti-Castro Cuban exiles landed at the Bay of Pigs as the first step in an invasion of Cuba. The scheme had been formulated under Eisenhower, but was approved and authorized by JFK. Yet it was ill-conceived, and, at the last minute, JFK decided not to involve U.S. forces directly. He refused to supply air support to the invaders, thereby dooming their mission to failure. Castro crushed the invasion within 48 hours, and the affair was a major embarrassment to the United States.

The following year, in October 1962, U.S. reconnaissance flights over Cuba discovered that the Soviets were building nuclear missile bases on the island. Although many of his advisers urged JFK to respond by attacking Cuba in force, the president wisely feared provoking a major war—perhaps even a nuclear holocaust. Instead, he set up a naval and air blockade of Cuba, demanding the removal of the missiles from the island, and announcing the U.S. intention to inspect and turn back any vessels carrying weapons. The world indeed seemed poised on the brink of nuclear war, but Soviet premier Nikita Krushchev at last agreed to withdraw the missiles in exchange for JFK's pledge not to invade Cuba and a further promise to remove U.S. missiles from bases in Turkey.

Assassination

The successful resolution of the Cuban Missile Crisis not only appeared to redeem JFK from the Bay of Pigs fiasco, it showed him as a strong, yet prudent leader, and it even improved U.S.-Soviet relations somewhat—at least to the extent that the two leaders agreed to communicate openly in order to avert future crises (a "hot line" was installed in the White House and the Kremlin, providing a direct communications link between Kennedy and Krushchev). By 1963, the two nations even concluded the Nuclear Test Ban Treaty, by which both nations agreed to cease atmospheric testing of nuclear weapons. It was an important first step toward mutual arms reduction.

Late in 1963, JFK seemed headed for a second term as president, when, on November 22, during a trip to Dallas, Texas, his motorcade was fired upon, and he was fatally wounded. (Texas governor John Connolly was also gravely wounded.) A lone assassin, Lee Harvey Oswald, was quickly identified and apprehended, but was himself murdered by one Jack Ruby, a shady Dallas nightclub owner with connections to organized crime, while he was being transferred from Dallas police headquarters. The assassination, a great national trauma, has remained a subject of intense national fascination, shrouded in a host of sinister conspiracy theories, some of which implicate government officials at the highest levels.

LBJ and the Great Society

JFK's vice president, Lyndon B. Johnson, pledged to honor Kennedy's legacy by carrying out his social and other programs. Under LBJ, Kennedy's New Frontier became the "Great Society," and, indeed, LBJ used Kennedy's "martyrdom" to supply the necessary momentum in Congress to achieve passage of a wide range of social programs. The legislation may have originated during the JFK administration, but it was Johnson who not only expanded the programs, but achieved their passage during the balance of JFK's term and after soundly defeating conservative Republican candidate Barry Goldwater in 1964. Under LBJ:

- A "domestic Peace Corps," Volunteers in Service to America (VISTA), was a hands-on aid program to the nation's poorest communities.
- The 24th Amendment was ratified, outlawing the poll taxes by which some Southern states had discouraged poor blacks from voting.
- The Civil Rights Act of 1964 was the most sweeping civil rights legislation since the end of the Civil War. It prohibited discrimination on the basis of race, religion, national origin, or sex in employment, education, and in access to all public accommodations.
- The Economic Opportunity Act set up the Office of Economic Opportunity (OEO), which administered Operation Head Start (a preschool program), Job Corps (to help high school dropouts gain employment), Neighborhood Youth Corps (for unemployed teens), Upward Bound (college tuition aid program), and other programs.

- Social Security was expanded.
- The Department of Housing and Urban Development was created to address the decay of the cities.
- New and extensive environmental legislation was enacted.
- Medicare and Medicaid programs were created for the elderly.
- The National Endowment for the Arts and the National Endowment for the Humanities were established to foster the arts and culture.
- The Voting Rights Act ensured enfranchisement for Southern blacks.
- The Immigration Act set a limit on annual immigration, but also equalized immigration quotas among Europe, Asia, and Africa.
- The Civil Rights Act of 1968 expanded the 1964 act by prohibiting housing discrimination.

The Vietnam War

LBJ clearly had a passion for social justice and a desire to help the poor and disadvantaged. However, his Great Society was ultimately compromised by his growing commitment to the war in Vietnam, which sapped funding resources from the social programs and which, ultimately, resulted in the fourth bloodiest conflict in U.S. history.

U.S. involvement in Vietnam began under Eisenhower, whose "domino theory" proposed support for the ostensibly democratic government of South Vietnam against takeover by the Communist North, lest, as a result of the fall of the South, the other nations of southeast Asia would fall to Communism like so many dominoes. Under JFK, U.S. involvement in the war increased: 15,500 military "advisors" were sent to the country. In 1964, President Johnson secured congressional passage of the Gulf of Tonkin Resolution following an unprovoked attack by North Vietnamese torpedo boats on two U.S. destroyers in the Gulf of Tonkin. The resolution gave Johnson extraordinary powers to commit troops and other assets to the conflict. (Doubts were later raised about the authenticity of the attacks on the destroyers.) By 1965, 75,000 Americans were fighting in Vietnam; by 1966, 375,000; by the end of Johnson's term, over half a million—many of the troops unwilling draftees.

By 1966, on the homefront, opposition to the war was mounting, especially as it became clear that the South Vietnamese regime the United States supported was not only poorly committed to the struggle, but was, in fact, corrupt and anything but truly democratic. By 1968, antiwar protest had polarized American public opinion, often pitting those of draft age against their elders. Johnson, faced with bitter opposition to his policies, withdrew from the upcoming presidential election, and Richard Nixon (whose political career many had prematurely written off after his defeat by JFK) defeated Democratic candidate Hubert H. Humphrey. Nixon, determined to turn over most of the fighting to the poorly motivated South Vietnamese army, tried to buy time for the American withdrawal by illegally bombing North Vietnamese supply lines and bases in Cambodia. This sparked massive antiwar demonstrations across U.S. college campuses, including one at Kent State University, which resulted in the deaths of four students from National Guard bullets (May 4, 1970). In response to this tragedy and to revelations (in "The Pentagon Papers") of a longterm secret government policy of deception and deceit in Vietnam, Congress cut off funds for the war, prompting Nixon and his negotiator Henry Kissinger to settle with the North Vietnamese in the Paris Peace Accords of 1973. Within two years, North Vietnam had totally defeated the South and reunited Vietnam under a Communist government. Perhaps no experience since the Civil War was more divisive and bitter to Americans than the Vietnam War.

CHAPTER 14

RECENT HISTORY (SINCE 1968)

THE STUDY OF RECENT AND CONTEMPORARY HISTORY IS BOTH EASIER AND MORE DIFficult than the study of earlier history. It is easier, because the materials of that history are fresher, and, indeed, many of the participants in and witnesses to that history may still be alive and available. It is also more difficult because insufficient time may have passed to allow judgments to be formed concerning the significance of certain events. In any case, it is in their coverage of recent history that introductory-level college and university courses are bound to vary most. What follows are some themes to expect.

Space Race Culmination: Moon Landing

For all the turbulence, disharmony, and dissatisfaction of the late 1960s, the nation and the world were united in wonder and awe when, on July 20, 1969, at 4:17 p.m. (Eastern Daylight Time), the lunar excursion module (LEM) *Eagle* landed two American astronauts, Neil Armstrong and "Buzz" Aldrin, on the moon. It was a national and human triumph that came at a time when a cynical nation and world needed it most. It also signaled U.S. victory in the "space race" against the Soviets, greatly adding to national prestige and testifying to the achievement of American technology and the culture that had produced it.

Culture, Counterculture, and Drug Culture

Since the early twentieth century, illegal drug abuse had been associated with the fringes of society, including (to some extent) with urban African Americans. By the 1950s, addiction to such narcotics as heroin was becoming a major problem in many American cities and was linked to the increasing incidence of violent street crime. In the 1960s, a new generation of middle-class white youth, characterized by relative affluence and the advantages of education, became passionately dedicated to forms of music and other types of popular art that expressed a turning away from much that had been accepted as the "American dream": material prosperity, a successful career, a happy marriage, and so on. The culture of protest included widespread use of so-called recreational drugs, including marijuana and the hallucinogenic LSD.

In addition to rock music and drugs, other forms of "countercultural" expression (or protest) included extravagant clothing and hairstyles, promiscuous sexuality, and the embrace of non-Judeo-Christian religious traditions, especially forms of Hindu belief. Those who had, to one degree or another, "turned on, tuned in, and dropped out" (to use the phrase of LSD advocate and former Harvard professor Dr. Timothy Leary) characteristically called themselves *hippies* (derived from *hip*, slang for being attuned to the latest social trends); and the hippie movement, despite its association with drug-induced escapism, was certainly not all negative. There was an emphasis on kindness, on affection, on looking out for one's fellow being, on caring for the natural environment, on social justice, on freedom of expression, on tolerance, on fostering creativity, on general peaceful coexistence, and on other life-affirming values.

For many, the era and the hippie experience was summed up in the summer of 1969, called the "summer of love" and capped by an open-air rock-music festival held on a farm near Woodstock, New York, during August 15-17, 1969. The Woodstock festival immediately became the symbol of a generation's solidarity in rebellion against the "Establishment" (as those who controlled the status quo were collectively labeled) and its war in Vietnam.

Nixon and China

For hippies and others with liberal inclination, the administration of Richard Nixon was a reactionary enemy clinging to outworn and destructive social and political values. Yet it was Nixon, a man who had forged his career beginning in the McCarthy era as a hardline anti-Communist, who reached out to normalize relations with Communist China and the Soviet Union.

In 1968, the United Nations sponsored the Nuclear Non-Proliferation Treaty, which sought to limit the spread of nuclear weapons by persuading those nations without nuclear arsenals to renounce acquiring them in return for a pledge from the nuclear powers that they would reduce the size of their stockpiles. The following year, under Nixon, the United States began negotiations with the Soviet Union to limit strategic (that is, nuclear-armed) forces. These Strategic Arms Limitation Talks—known by the acronym SALT —produced a pair of important arms-control agreements in 1972. From 1972 to 1979, SALT II was conducted, extending and more precisely defining provisions formulated in 1972. Although the U.S. Senate failed to ratify the treaty, the two nations generally abided by its arms-limitation and arms-reduction provisions.

Perhaps even more remarkable was President Nixon's February 1972 journey to China, where he was received in Beijing by Chairman Mao Tse-tung, the very incarnation of the communism Nixon had spent his life opposing. Nixon reversed the long-standing U.S. policy of refusing to recognize China's communist government, and, by January 1979 (under President Jimmy Carter), full diplomatic relations were established between the nations.

Through his emissary Henry Kissinger, President Nixon was also instrumental in negotiating a cease-fire and a degree of lasting peace between the warring factions of the Middle East.

The Watergate Crisis

While Nixon had proven himself a bold visionary on the international front, he failed to win trust and confidence at home. Nevertheless, there was little doubt that Nixon would be reelected in 1972. Yet Nixon sought a sure thing, desiring not merely to defeat the Democrats, but to crush them. Accordingly, he authorized a campaign of espionage against the Democratic party and a program of "dirty tricks" aimed at smearing Democratic challengers. On June 17, 1972, during the presidential campaign, five burglars were arrested in the headquarters of the Democratic National Committee at the Watergate office building in Washington, D.C. They included three anti-Castro Cuban refugees, all veterans of the ill-fated Bay of Pigs invasion, and James McCord, Jr., former CIA agent and now "security" officer for Nixon's reelection committee. Through months of investigative reporting by Washington Post reporters Bob Woodward and Carl Bernstein, the burglary was traced to the White House and, ultimately,

Richard Nixon, his top aides, and advisors.

Despite arrests and early revelations, President Nixon won reelection, but, soon after he began his second term, the Watergate conspiracy rapidly unraveled. One after another, the "president's men" gave testimony to federal authorities, and, in February 1973, the Senate commenced the televised Watergate Hearings. After each key disclosure, the president announced the resignation of an important aide, including John Ehrlichman and H. R. Haldeman, his closest advisors. Nixon's counsel, John W. Dean III, was dismissed. It became increasingly clear that the president himself had grossly abused his office and had used government agencies to attack his political enemies and then to obstruct the Watergate investigation itself. In the midst of all this, Vice President Spiro T. Agnew was indicted for bribes he had taken as Maryland governor. He resigned as vice president in October 1973 and was replaced by Congressman Gerald Ford of Michigan. At last, it was revealed that President Nixon had covertly taped White House conversations; the tapes were subpoenaed, but the president claimed "executive privilege" and withheld them. Ultimately, the tapes were released, and, on July 27-30, the House Judiciary Committee recommended that Nixon be impeached on charges of obstruction of justice, abuse of presidential powers, and attempting to impede the impeachment process by defying committee subpoenas. On August 9, 1974, Nixon became the first president in U.S. history to resign from office.

OPEC and the Energy Crisis

Gerald Ford served out the remainder of Nixon's term, but lost the 1976 election to Democrat Jimmy Carter. Carter inherited a nation whose economy had entered a period of "stagflation," a condition characterized by high prices *and* low demand—characteristics of recession as well as inflation. This depressed and demoralized the nation, which then suffered an energy crisis brought about when the Organization of Petroleum Exporting Countries (OPEC), a Middle Eastern oil-producing cartel, imposed in 1979 its second embargo of the decade. (The first, late in 1973, was imposed on the United States and other nations that supported Israel in its 1973 war with Egypt.)

The energy crisis meant high costs for gasoline and heating oil, and it also meant gasoline shortages, which impinged upon Americans' long-cherished "right" to drive anywhere anytime. The energy crisis also aggravated stagflation. To many Americans during this period, the United States seemed to have become an impotent giant, and President Carter tended to bear the brunt of his fellow citizens' bad feelings.

In the meantime, a revolution in Iran, led by the Ayatollah Ruhollah Khomeini, toppled longtime U.S. ally Muhammad Reza Shah Pahlavi, the Shah of Iran, who fled into exile in January 1979. In October, desperately ill with cancer, the Shah was granted permission to come to the United States for medical treatment. In response to this gesture, on November 4, 500 Iranians stormed the U.S. embassy in Tehran and took 66 U.S. embassy employees hostage, demanding the return of the Shah.

A few hostages were released, but 53 remained in captivity for 444 days, the remainder of Carter's term. An attempt to rescue them proved abortive, and, much as the nation's economic woes were laid at the doorstep of the White House, so was the inability to safeguard the embassy personnel. In the elections of 1980, Carter suffered a decisive defeat to Ronald Reagan, a former B-picture actor and popular governor of California. The hostages were released on the day of the new president's inauguration.

Reaganomics

On the domestic front, President Nixon had developed what he called the "New Federalism," cutting back on federal programs and shifting power as well as responsibility back onto state and local governments. President Carter had reversed this trend somewhat—though he was limited by the constraints imposed by the ailing economy—but now President Reagan embraced it again. Expressing admiration for the hands-off presidency of Calvin Coolidge, Reagan pledged to shrink government and withdraw it from many sectors of national life.

He also introduced a "new" economic policy, intended to "jump start" the stagflated economy. Dubbed "Reaganomics," it was based on the president's assertion that "government is the problem" as far as the economy was concerned. During the Reagan administration, many of the social programs of Democratic administrations were drastically cut, defense spending was dramatically increased, and personal income taxes were reduced. President Reagan's economic strategists held that, despite reductions in aid to the poor and disadvantaged, the tax cuts and economic incentives at the upper end of economic pyramid (the "supply side") would "trickle down" to all levels of the economy and society. Indeed, the economy was "jump started" in that the 1980s were characterized by an avalanche of corporate acquisitions and mergers (as well as Wall Street scandals); however, the national debt soared, and, on October 19, 1987, the stock market plunged 508 points (almost double the fall of the 1929 crash). While the market gradually recovered—and no depression ensued—the event seemed to mark the end of the Reaganomics experiment.

The Cold War Ends

Many Americans were dismayed by the Reagan administration's apparent indifference to the growing legion of "homeless people" who haunted the nation's large cities and even many of its smaller towns. The administration likewise turned away from the AIDS epidemic, which developed initially among gay (homosexual) men, but was soon also diagnosed in "straight" men, in women, and in children. The first AIDS cases were formally reported in 1981, but President Reagan failed even to make public mention of the disease until April 1987, fully six years after health officials had determined that a major epidemic was under way.

But if the Reagan administration did not engage AIDS vigorously, it did not hesitate to take on the Soviet Union, assuming an aggressive stance against what the president called "an evil empire." Defense spending was dramatically increased, dwarfing domestic budget cuts in welfare and other programs.

The president also acted aggressively to meet perceived military threats throughout the world, sending U.S. marines in the summer of 1982 to Lebanon as a peacekeeping force. On October 23, 1983, more than two hundred of these troops were killed in their sleep when a truck laden with 25,000 pounds of TNT was driven into the marines' Beirut headquarters building. Just two days after this disaster, the president ordered an invasion of the island nation of Grenada in the West Indies. Cuban troops had been sent to the tiny country (population 110,100) at the behest of its anti-American dictatorship, and the president was determined to protect the approximately 1,000 American citizens there. The president also saw a successful "liberation" of the country as a kind of emotional compensation for the death of the marines in Beirut.

During 1983, President Reagan announced the most spectacular, ambitious, elaborate, and expensive military project in American history. It was called the Strategic Defense Initiative (SDI), but the popular press dubbed the system "Star Wars," after the popular George Lucas science fiction movie of 1977. Using an orbiting weapons system, the idea was to create a "shield" against ballistic missile attack by destroying incoming ICBMs before they began their descent. The weaponry, however, was far beyond the cutting edge of technology, and many believed the system could never be made to work. Moreover, critics saw it as a dangerous violation of the 1972 ABM (antiballistic missile) treaty: a temptation to thermonuclear war because it promised to make such a war survivable. Others pointed out that the $100 to $200 billion required to build SDI would permanently cripple the nation. (Indeed, after Presidents Reagan and George Bush had spent some $30 billion on SDI, it was revealed in 1993 that at least one major space test had been "rigged" to yield successful results, and Caspar Weinberger, Reagan's secretary of defense, suggested that the entire Star Wars program had been an elaborate decoy, designed solely to dupe the Soviet Union into spending a huge proportion of its resources on a Star Wars program of its own—a program that U.S. scientists already knew was unworkable.

Decoy or not, the United States did outspend the Soviet Union, which, during the 1980s, weakened and buckled. In 1987, standing near the Berlin Wall, the brick, stone, and razor-wire symbol of a half-century of Communist oppression, President Reagan made a speech calling out to Soviet leader Mikhail Gorbachev: "Open this gate! Mr. Gorbachev, tear down this wall!" Two years later, Berliners began chipping away at the wall, tearing it down piece by piece. The Soviets merely looked on. By the end of the 1980s, it was clear that the long Cold War had ended. Historians debate whether the policies of Ronald Reagan *won* the Cold War or whether the Soviet Union, shackled to an inherently flawed economic and political system, simply *lost.*

Iran-Contra Scandal

In November 1986, President Ronald Reagan confirmed reports that the United States had secretly sold arms to its implacable enemy, Iran. The president at first denied, however, that the purpose of the sale was to obtain the release of U.S. hostages held by terrorists in perpetually war-torn Lebanon, but later admitted an arms-for-hostages swap. Then the plot thickened—shockingly—when Attorney General Edwin Meese learned that a portion of the arms profits had been diverted to finance so-called Contra rebels fighting against the leftist Sandinista government of Nicaragua. As part of the ongoing U.S. policy of containing communism—the policy begun by Harry S Truman—the Reagan administration supported right-wing rebellion in Nicaragua, but Congress specifically prohibited aid to the Contras. The *secret* diversion of the *secret* arms profits was blatantly unconstitutional and illegal. A lengthy investigation revealed that a cabal of Israelis had approached National Security Advisor Robert MacFarlane with a scheme in which Iran would use its influence to free the U.S. hostages held in Lebanon in exchange for arms. Secretary of State George Schultz and Secretary of Defense Caspar Weinberger objected to the plan, but (MacFarlane testified) President Reagan agreed to it. In a bizarre twist, U.S. Marine lieutenant colonel Oliver ("Ollie") North then modified the scheme in order to funnel profits from the arms sales to the Contras.

Desert Storm

Ultimately, few people believed that President Reagan had been ignorant of the scheme, but even if he had been the unwitting dupe of zealots in his administration, the implications were even worse, painting a picture of a passive chief executive blindly delegating authority to his staff. Despite the indictment and even conviction of some White House staffers (including North), no one was punished in the affair.

While Iran-Contra and the 1987 stock market crash marred Reagan's second term, his vice president, George Bush, sailed to an easy victory over Democratic contender Michael Dukakis in 1988. But Bush lacked Reagan's charisma, and, in the face of a continually faltering economy, he seemed doomed to a one-term presidency—until, on August 2, 1990, Iraqi president Saddam Hussein ordered an invasion of the small, oil-rich Arab state of Kuwait. In this tense and dramatic crisis, the Bush administration brilliantly used the United Nations to sanction action against Iraq, and, with masterful diplomacy, assembled an unprecedented coalition of nations—Argentina, Australia, Bahrain, Bangladesh, Belgium, Canada, Czechoslovakia, Denmark, Egypt, France, Greece, Hungary, Italy, Kuwait, Morocco, the Netherlands, New Zealand, Niger, Norway, Oman, Pakistan, Poland, Qatar, Saudi Arabia, Senegal, South Korea, Spain, Syria, the United Arab Emirates, United Kingdom, plus the United States—to oppose the invasion. The United States committed the largest force since Vietnam: more than a half-million troops, 1,800 aircraft, and some 100 ships. Indeed, many Americans, lacking faith in the prowess of U.S. arms, feared that Iraq would be the start of another Vietnam. Instead, after Iraq failed to withdraw from Kuwait, a massive air and ground war was launched during January-February 1991. Combined U.S. and coalition forces in the Gulf War amounted to 530,000 troops opposed to 545,000 Iraqis. U.S. and coalition losses were 149 killed, 238 wounded, 81 missing, and 13 taken prisoner (they were subsequently released), whereas Iraqi losses have been estimated in excess of 80,000 men, with overwhelming loss of materiel.

New Directions

For President Bush, the afterglow of Desert Storm did not last long. Economic problems persisted, and, running on the theme of the economy (as well as appealing to an element of liberal conscience), Bill Clinton overcame sexual scandal to defeat Bush in his bid for a second term.

In 1996, Clinton was reelected to a second term. Themes of the Clinton years include:

- **Brief resurgence of third-party politics:** Texas billionaire H. Ross Perot received a stunning 19 percent of the popular vote—far more than any other third-party candidate (including Theodore Roosevelt) had ever received.
- **Strong polarization of the political right and left**
- **Resurgence of the American economy;** the stock market soared to record levels
- **Domestic terrorism:** the nineties saw a rash of apparently politically motivated bombings, including the World Trade Center in New York, a federal building in Oklahoma City, Olympic Centennial Park in Atlanta, and several abortion clinic bombings
- **The rise of interactive communications:** millions of Americans became linked to the vast Internet during the 1990s
- **Political and personal scandal:** President Clinton's second term was plagued by sexual scandal and the issue of impeachment on grounds of perjury. As the 1990s drew to a close, many Americans felt that their quality of life was good, but that the moral integrity of the nation's political leaders was a serious problem for the current state and future of the nation.

PART THREE

MIDTERMS AND FINALS

CHAPTER 15

BRADLEY UNIVERSITY

HISTORY 204: UNITED STATES HISTORY SINCE 1877

Timothy P. Maga, Oglesby Professor of American Heritage

LIKE ANY SURVEY COURSE IN ANY FIELD, HISTORY 204 IS THE FOUNDATION COURSE required before moving on to more specialized study. With that in mind, lively, engaging teaching is a necessity. For many, this course is their introduction to the entire field of history. It can determine whether they will take another history course or avoid the field altogether. Consequently, it is a very important course to the survival of any history department as well as to student academic pursuits. For the sake of the department, the field, and the students themselves, I try to transform what could be dry, fast-moving survey material into a memorable lecture/discussion experience. Furthermore, with business and government hiring more history majors and minors than ever before, it is essential, I believe, that these students have a decent background in analytical writing. Their ability to argue a position clearly is important to on-the-job success after graduation. Hence my stress on analytical essay exams versus the multiple-guess approach or even the short-answer format. For too long, the history course has been associated with the multiple choice/true-or-false exam, thereby keeping the student away from analytical thinking and more involved in rote memory efforts or guesswork. In today's global and highly competitive marketplace, the history student cannot afford to be lacking in analytical prowess. Examining historical dilemmas and controversies, many of which are relevant to the present, assists in building that prowess. Thus my primary objectives here are linked to interesting students in the historical field and profession as well as training them to think analytically through the examination of historical problems."

MIDTERM EXAM

Section A *(30 percent):*
Answer and explain the significance of THREE of the following.

(1) Espionage and Sedition Acts
(2) Teapot Dome Scandal
(3) Billy Mitchell and Pete Ellis
(4) Bonus March
(5) Cordell Hull

On the first day of class, I spend quite a while going through a brief, no-nonsense, but detailed syllabus with the students. The course requirements are made quite obvious from the beginning. With that in mind, I pass out exam reviews one week before each of the three exams in this course. The sample questions on that review sheet, students are told, are very similar to the ones that they will see on exam day itself; hence, they are given some time to prepare a thesis to those questions and also prepare to defend it with the wealth of course material (from books and lecture). In that defense, they also know that more is better than less (i.e., data must be explained if used to defend a thesis and that their mastery of course material must, at the same time, be demonstrated on paper during these in-class exams). I urge these students to take the review period most seriously. I urge them to e-mail me any questions, take advantage of my office hours for personal consultations, or walk in whenever it's convenient for all of us. In short, they know that I will go the distance to assist any serious student struggling with the age-old problem of building a thesis and defending it persuasively with a mountain of data. They also know that I publish a lot, won a Pulitzer Prize in history, and can truly help them out if they seek it.

ANSWERS

(1) The Espionage and Sedition Acts were modeled off of John Adams's 1798 Alien and Sedition Acts. The latter crushed dissent during the divisive debate over possible U.S. assistance to the French Revolution. Like this controversial period in the Adams presidency, the Espionage and Sedition Acts symbolize the presidency's (Woodrow Wilson) demand for conformity and solidarity in public opinion during times of "national emergency." Following the 1917 U.S. entry into World War I, President Wilson, a former history professor who admired President Adams's 1798 decision-making, acted in a sense of "executive privilege," claiming that his role of commander-in-chief justified any policy, as World War I continued, which protected national security. This gave him the power, he argued, to crush any and all opposition to World War I whenever he saw fit. The Supreme Court agreed. Although jousting against First Amendment rights, the Supreme Court of Chief Justice Oliver Wendell Holmes upheld presidential power over civil rights/liberties protection. Wilson hailed this development as a great victory, and it still has relevance to the presidency today; however, it remained a politically unpopular decision in Wilson' s day. Given that unpopularity, the Acts even contributed to the defeat of Wilson' s Democratic Party in the 1920 election; however, the question of whether civil rights/civil liberties can survive in a democracy at war remains a matter of controversy and doubt.

Depending on writing style, if the student provides an explanation of the 'significance' in history of the Espionage and Sedition Acts, a brief account is more than possible. Noting that significance, combined with some detail or background, equals the mini-analysis that is required in this opening section of the exam.

The answer above represents a concise, but thoroughly acceptable account of the Acts and their impact. Further detail on the Holmes case itself, Wilson's career and sense of mission in foreign policy making, his commitment to creating a certain First World status for American power, etc., is also possible here. Indeed, these types of issues will often appear in this ientification-type response as well, and, since they shed light on the 'big picture' noted above, they are also thoroughly acceptable. In short, a combination of the hardcore, explanatory facts, accompanied by some comment on their historical significance and impact, equals success here. The goal remains a mini-essay or smaller version of the Section B (long-essay) effort, but the commitment to analysis must remain the same.

The Espionage and Sedition Acts exercise was selected by nearly 90 percent of the students who took the exam. Their reasons for this decision varied from interest in the Acts and resulting court case which tested them, to a conclusion that the data associated with this exercise was "easier to manage" than other selections.

(3) Billy Mitchell and Pete Ellis: Following the Washington Conference of 1921-22, U.S. defense policy stressed disarmament and peace over international obligation and Wilsonian "collective security." Ironically, the U.S. military was assigned the task to destroy its own best hardware in the name of the Harding and Coolidge administration's commitment to isolationism. One of those military figures in charge of this destruction, Gen. Billy Mitchell, a pioneer aviator and brilliant tactician, hoped this otherwise unfortunate experience for the professional soldier held hidden opportunities. Although he faced many political obstacles, Mitchell demonstrated that the U.S. military could be small, professional, and victorious if it stressed aircraft carriers over cruisers and destroyers. In fact, in one dramatic demonstration staged by Mitchell, America's one aircraft carrier at the time, *USS Enterprise*, sunk several WW I-era cruisers and destroyers with planes from its own decks.

Hence, he predicted that all future naval battles would be won by carriers, and he also predicted that America's next and most dangerous rival was the empire of Japan. Both traditional U.S. navy strategists and Japanese government officials demanded that action be taken against Mitchell for his "reckless and damaging" predictions, and he was court-martialed for his efforts. He was later considered a hero and a victim to many who saw him as a genius "ahead of his time."

Col. Pete Ellis, USMC, the founding father of modern U.S. Navy and Marine Corps Intelligence, was a staunch early 1920s supporter of Mitchell, who tried to continue Mitchell's work. His personally led espionage missions to Japan and the Japanese-held islands of the Western Pacific proved Japan's militarist intentions as early as 1924. This temporarily won him the support of President Coolidge, who then devised a contingency defense plan against Japan; however, Ellis, like Mitchell, was opposed by many of his own, more conventional colleagues.

Ellis's further experiments with his personally designed amphibious landing craft or LST were labeled too expensive and wasteful by those colleagues, and, like Mitchell, he would have faced court-martial if he had not been killed in an espionage mission to the Palau islands in the mid-1920s.

Both Mitchell and Ellis represent the military's best effort to maintain readiness and effectiveness in the darkest days of U.S. isolationism.

The approach used in response to the Espionage and Sedition Acts question can be applied to all of questions in Section A of the midterm. For instance, the basic facts surrounding #4 (the Bonus March) or the contributions of America' s longest-reigning secretary of state, Cordell Hull (#5) are to be expected in a successful student answer. This, plus some comment on significance, impact, influence on past, present, or even the future, remains part of the assignment as well.

My students' greatest difficulty lies in determining the meaning and significance of events. For example, in the Espionage and Sedition Acts, I may only see comments in reference to the fact that they existed, and little else. No comment about their impact on 'executive privilege' or even Wilson' s political fortunes. Noting a fact or two without considering the significance of those facts is an unacceptable answer in a Section A exercise.

(5) Cordell Hull: Cordell Hull holds the record as the longest-serving secretary of state in the history of the State Department. Although he described himself as "a poor country lawyer," Hull was one of the South's most respected judges before turning to diplomatic work. He opposed the isolationist conferences of his predecessors at the State Department, and Franklin Roosevelt brought him into the original "Brain Trust" of his New Deal administration because of his ideas on restoring an activist economic foreign policy. Nicknamed "reciprocal trade," this policy was meant to restore a free-flowing trade relationship with America's trade partners who had been lost in the era of the Smoot-Hawley tariff. The latter stimulated a trade war, which destroyed free-trade relationships. Hull was most proud of his trade policy efforts, noting that a good foreign policy should also have a healthy impact on pocketbook issues domestically. He would be less proud of his World War II-era diplomacy, for President Roosevelt took a strong personal role in wartime policy-making. He retired in 1944, believing that the State Department was not as powerful and influential in foreign policy-making as it had been one decade earlier.

Thus Hull represents a transition between the more independent secretaries of state of an earlier era versus the postwar period of "executive privilege," the National Security Council, and the Central Intelligence Agency.

Section B *(70 percent):*

Select ONE of the following. State your thesis clearly and be precise.

(1) Although they maintained a different style and approach, Woodrow Wilson and Franklin Roosevelt would have agreed that the president must be a powerful and active one. Warren Harding, Calvin Coolidge, and Herbert Hoover believed that American power should be defined by the private sector, the states, and, at times, the U.S. Congress. Which point of view, as you see it, offered the most effective government? Why or why not?

(2) From the end of World War I to Pearl Harbor, most Americans believed that the best foreign policy is none at all. In the late 1990s, this is considered foolish. To 1920s and '30s Americans, it made perfect sense. Why? Examine the isolationist vs. interventionist debate of this interwar era.

The question asked implies and suggests to the student that a comparison-contrast of 1990s versus 1920s-'30s attitudes towards U.S. influence in the world would help establish a thesis. Students get a very comprehensive review from me beginning one week before the exam date. Hence, most of them have an idea what to expect on Section B, and the decent students come to the test with at least a good idea of how they want to answer key questions. In any event, the comparison-contrast model is not required. The better students will detail and examine the matter at hand without wasting precious writing time or overly general rhetoric on the comparison-contrast model. With this in mind, the following answer stresses a straightforward, no-nonsense approach to the question.

ANSWER

To President Woodrow Wilson and his Progressive "interventionist" supporters in the First World War, America was destined to "make the world safe for democracy," and the world's leading democracy, America, would assure that World War I "was the war to end all the wars." To accomplish this dream of world power influence and clout, America needed, said Wilson, to maintain a military presence in Europe and elsewhere, remain ready to crush monarchy or totalitarianism at a moment's notice, fight good fights such as world decolonization, and aid all who share in the democratic promise. To most Americans, this messianic agenda was not their own. Most wanted their sons and husbands home from the First World War, and most sought peace and economic security after a horrible conflict. Wilson and the Progressives were powerless to influence this understandable position and emotion. Unfortunately for them, they also did not respect it, and they paid the price (namely Wilson's Democratic Party) at the polls in 1920.

Promising "normalcy," the new Republican and ultraconservative president, Warren Harding, also promised disarmament instead of "collective security," diplomacy instead of sword-rattling, and a commitment to domestic development over the schemes and dreams of global interventionists. Harding even blamed activism in the White House for America's troubles, noting that power should rest in the business community and, in general, outside of Washington, D.C. Sadly, for this point of view, Harding's presidency was associated with horrible corruption, ranging from questions about his commitment to prohibition to an effort to make millions of dollars from an illegal oil deal in Teapot Dome, Wyoming. The latter actually put the U.S. navy's oil reserves in peril in the name of greed, and during a time of diplomatic wrangling with the government of Japan over Western Pacific island affairs (the Yap Crisis).

Although divorced from corruption, President Calvin Coolidge continued the isolationist agenda, inheriting the conclusions of the 1921-22 Washington disarmament conference following President Harding's 1923 death in office. That conference led to the destruction of America's World War I-built military might and hardware to the detriment of an adequate defense (especially in the Pacific). Coolidge would commission contingency studies over Pacific military crises (Orange-Blue Plans), but this was largely a paper exercise, for the disarmament campaign continued. According to Secretary of State Charles Evans Hughes, the destruction of the U.S. military would lead to defense expenditure invested in needed domestic projects (as well as encourage peace abroad). Hence he sloganed that "disarmament plus economy equals peace." Unfortunately for Hughes, he created something of the Frankenstein's monster. When he proposed a delegate or even just a U.S. observer to the World Court in The Hague (in order to argue America's case, for instance, if U.S. shipping was abused on the high seas), Hughes was accused by fellow isolationist Republicans of having abandoned isolationist principle. A role in the World Court was not acceptable politics, and Hughes left his post largely because of this turn of events. His successor, Frank Kellogg, would attach his name to a symbol of 1920s isolationist policy, the Kellogg-Briand Pact. A simple document, co-signed by the French prime minister and foreign minister, Aristide Briand, the Pact asked world governments to "outlaw war as an instrument of national policy." Ironically, Briand hoped this U.S.- and French-led effort would win America's support for various French military defense plans in Europe. Meanwhile, the document would soon be signed by fascist leaders, such as Mussolini and Hitler, as well as Joseph Stalin of the Soviet Union. Hence, it was more of a dream for peace than enforceable policy, and America would soon look foolish for committing itself to that dream above its own adequate defense.

Herbert Hoover continued the isolationist tradition, and, given his pacifist Quaker roots, added a certain religious dedication to the commitment to isolationist peace. This would be especially proven in the 1931-32 Japanese invasion of Manchuria and resulting daylight bombing raids (the first ever) over civilian targets in Shanghai. Hoover refused to "recognize" Japanese aggression and created a policy of "non-recognition" called the Hoover-Stimson Doctrine; consequently, totalitarian governments, such as Prince Konoye's Japan, soon viewed America as so weak and escapist that it would not even aid a longtime ally, such as China. Thus isolationism, many believe, encouraged totalitarian leaders in both Asia and Europe to take advantage of America's retreat.

With the rise of Franklin Roosevelt and the Democrats in 1932, isolationism did not disappear simply because the new president questioned the logic of isolation. The U.S. Congress, dominated by Roosevelt's own party, passed three anti-presidential and punitive acts (the Neutrality Acts between 1935 and 1937). All three attempted to halt any presidential action that would lead to an aggressive U.S. posture against totalitarianism (namely the Italian invasion of Ethiopia, Hitler's invasion of the Rhineland, and the Spanish Civil War). A fourth act was passed in 1939, but it largely confirmed what already existed.

Roosevelt tested isolationist opinion through an angry, but controlled response (in a public speech of Oct. 1937) to the Japanese invasion of China proper. Calling for a "quarantine of aggression," some thought Roosevelt was asking for a declaration of war against Japan. This was not the case, but some, such as *The Chicago Tribune*, even asked for Roosevelt's impeachment for having violated the "spirit" of the three neutrality acts. Since one cannot be impeached for violating a "spirit," Roosevelt was safe from prosecution (but worried about the blindness of isolationism). His efforts to defeat a bill by fellow Democrat and Congressman Louis Ludlow that would further legalize isloationism as the nation's foreign policy were barely successful, but at least Roosevelt could claim that his ideas on rebuilding U.S. defense were starting to get through to public opinion. Only the attack on Pearl Harbor, however, would halt the progress of isolationism and even destroy the popular pro-isolationist group America First, led by hero-aviator Charles Lindbergh.

In its day, isolationism reflected America's confusion over its place in the world. It had moved from a quiet farming nation in the era of the turn of the century to world power status in a few short years. Defining the responsibilities of that status suggested to many that the U.S. would become a "policeman" to the world, threatening endless war, and distracting the country from domestic priorities. Only when America's very being was threatened and injured by the fascist coalition (Rome-Berlin-Tokyo) and the resulting attacks by Japan on Pearl Harbor, Guam, and the Philippines did U.S. opinion change.

Students are given an exam review one week in advance of the test. On that exam review sheet, there are two Section B (long-essay) questions very similar to the actual exam questions. They know that they will be similar, and they are urged to prepare their "favorite" selection accordingly. Question 2 (answered here) was the more popular of the two questions and was selected by over 80 percent of the class. Some like the question because of its more "mechanical," cut-and-dry nature (student-labeled as a "good thing"), in contrast to the "philosophical" nature (a "bad thing") of Question 1.

To succeed with a question on the viability and popularity of U.S. isolationism in the 1920s and '30s, the student must first examine the public rejection of President Wilson' s Versailles conference commitments. Wilson' s mission to make the United States a world power with global military and economic responsibilities was rejected by majority public opinion, although Wilson refused to accept the rejection. Once the student examines the ending days of Wilsonian interventionism, making brief comments on the U.S. intervention in North Russia/Siberia and the threat of war with Japan over the tiny Pacific island of Yap, he may move on to the election of 1920, the landslide defeat of the Democrats, and the rise of Warren Harding, Secretary of State Charles Evans Hughes, and the new isolationist mission. The late 1921 Washington conference symbolizes the new regime's ambition in that regard, whereby Harding and Hughes insist on a 5-5-3 naval power ratio for the Big Three navies (U.S., Great Britain, and Japan). Hughes's interest in diplomacy via international law-making versus the sword-rattling style diplomacy of Theodore Roosevelt, William Howard Taft, and Woodrow Wilson merits comment, as does Hughes's hasty retirement due to his own legal interest in placing a delegate in the World Court. The latter episode was seen in public opinion as a contradiction, for one cannot be truly isolationist if you seek a delegate in the international World Court.

A discussion of the goals of Hughes's successor, Frank Kellogg, would illustrate the excesses of isolationism quite well, i.e., the Kellogg-Briand Pact especially. This development, bracketed by a brief discussion of the lack of personal interest on the part of Presidents Harding, Coolidge, and Hoover in the mechanics of foreign policy making (outside of limited adventures in economic foreign policy, such as the Dawes Plan) would then make sense here. Indeed, the presidency sought a certain retreat even from crisis management in foreign policy making, and President Herbert Hoover's reaction (or non-reaction) to the 1931 Japanese invasion of Manchuria proves the point.

In contrast to some opinions, isolationism did not disappear with Herbert Hoover and the rise of FDR's New Deal in 1933. It lived on, especially in Congress, where strong Neutrality Acts were passed before Pearl Harbor. Discussing each one of them keeps the students in a decent chronological line to the U.S. entry in World War II, thereby leading to their own conclusions on the impact and significance of the isolationist era.

Indeed, a strict chronology is encouraged of the students here. It will help keep the wealth of course material organized and workable for them. A topical approach, they are told, is better suited to the skilled analytical essay writer, such as the history graduate student. In the survey, they are just learning to develop a good thesis, argue it forward in a detailed, chronological

fashion, and make conclusions from the mountain of available data. The biggest problem faced here is reluctance to explain events (such as a Washington conference or a Kellogg-Briand Pact). Many students would rather list things than examine them. That problem is followed by those who are afraid, for one reason or another, to take a position on the question asked. I try to assure them that there is no "politically correct" answer, and that they should draw logical conclusions from the data that they have taken great pains to explain.

FINAL EXAM

Each of my three exams is weighted equally in terms of grading significance to the students. A small percentage is set aside for class participation, but the latter can, at the best of times, only positively influence a student, who, for instance, might be dangling between a B+ and an A-. That talkative student, if the talk was relevant, would then, of course, win the A-.

Section A *(30 percent):*
Answer and explain the significance of THREE of the following:

(1) Taft-Hartley Act
(2) *Brown v. Topeka Board of Education*
(3) Bay of Pigs and Missile Crisis
(4) Selma March
(5) Campaign Reform Act (1974)

ANSWERS

(1) The Taft-Hartley Act: The Taft-Hartley Act carries political, economic, and legal significance from the era of President Harry Truman to Bill Clinton. Originally sponsored by Senator Robert Taft, son of the late President William Howard Taft, the act was meant to protect both America's national security and its economic fortunes from a crippling strike. Taft reasoned that the late 1940s steel strike, given its crushing impact on the economy, was the equivalent of a wartime national security crisis. Indeed, he suggested that America's diehard Cold War opponent, the Soviet Union, benefitted from America' s labor crisis and resulting confusion, economic dislocation, and political deadlock. The act granted the president special "executive privilege" powers to break a strike as a means to defend the national economy and the country. If Truman invoked the act, his strongest political supporters, organized labor, might abandon the

Democratic Party because of it. If Truman did nothing, the strike might continue and further weaken a once seemingly unassailable post-World War II economy. To the Republican Taft, a presidential aspirant, the act put the Democratic Truman in a no-win situation, which could only benefit the president's opposition (or so he hoped). Ironically, the act also equaled a surrender of Congressional influence in labor/economic matters to the White House—a situation many Congressmen would later regret. Meanwhile, Truman did indeed sign the bill into law, but also vowed never to enforce it. The very decision to sign this controversial legislation helped stimulate the bargaining process and bring about the end of the strike. From this point forward, many Democrats running for president, preferring to court rather than annoy Big Labor, also vowed never to enforce Taft-Hartley. The first president to invoke it would be the Republican Ronald Reagan in 1981. Otherwise, presidential power had increased another dramatic step forward courtesy of the Taft-Hartley Act.

Because of its complexity and impact in a variety of areas, including politics, economics, labor history, Cold War decision-making, etc., the Taft-Hartley Act takes quite a bit of explaining by students. After having written two previous essay exams in this course (there are two midterm-level exams), the students, by this final exam, should be ready for the challenge. My own data proves the point, because for the vast majority of survey students select this item for a mini-analytical essay. Success here is determined by how well the students can weave this involved tale of political ambition, economic turmoil, real or imagined national security threats, and "executive privilege." The latter concept is one that the students discussed in the previous exam. Hence, Taft-Hartley could be considered a continuation of the tale of growing presidential powers. Some will even note that fact, drawing analytical connections to related "executive privilege" issues discussed in class or in the readings. Making those connections, as well as sorting through the act's complexities and its significance to later presidencies, adds to the student success story here. Again, a decent grade for their Taft-Hartley mini-essay requires the student to integrate the facts of the act with its significance and impact over the years. Throwing raw facts on paper is not good enough. Waxing poetic on overly general points is not acceptable, either. An analytical essay remains the goal, and most serious students, by this late date in the course, have the experience to succeed. As in a midterm exam, a commitment to analysis is expected in all three items selected. Section A of the exam can even serve as a "warm up" to the larger, more elaborate analytical tasks of Section B.

(3) Bay of Pigs and Missile Crisis: Following the rise to power of Cuba's Fidel Castro, CIA director Allen Dulles proposed to the outgoing Eisenhower administration a "special action" plan to eliminate this potential ally of the Soviet Union. Influenced by the strident anti-Communism of the day, Dulles and the Eisenhower cabinet also saw Castro as a potential agent for Soviet causes in Latin America. Castro also symbolized the lack of anti-Communist clout for the United States in the Western Hemisphere. The "special action" plan called for an invasion of Cuba led by CIA-trained Cuban exiles. Both Eisenhower and his successor, John Kennedy, approved it, even though the latter had made many glowing speeches about how the United States should warm to the Third World rather than invade it. Overly optimistic, the invasion landed near the beach of Castro's largest military base at the Bay of Pigs. This, it was believed in Washington, would make it easier for Castro to surrender. Instead, in 48 hours, the invasion was defeated, and Kennedy ruled out U.S. Air Force support in an effort to rescue it. The entire episode pushed Castro closer to the Soviet Union in the name of national defense and protection, and, by summer 1962, he won the Kremlin's support for medium- range nuclear missile support as well.

The U.S. discovery of these missiles in September 1962 by a passing U2 spy plane led to the Cuban Missile Crisis. Again, influenced by the rabid anti-communism of the day, the Kennedy cabinet assumed that the nation would not accept Cuba as a hostile threat to the U.S. Nuclear war was now possible, and 300 million dead was a projected casualty figure. All of Kennedy's advisors, with the exception of the UN Ambassador, Adlai Stevenson, advocated a quick U.S. nuclear strike on Castro. As the crisis worsened, the president's brother, the attorney general, Robert Kennedy, also moved toward a negotiated solution. After 13 days, an agreement was reached whereby U.S. Navy vessels, ordered to "quarantine" Soviet ships coming and going from Cuba, would not board and inspect those ships. In exchange, the Soviets promised to dismantle the missiles in Cuba. Precise detail was to be worked out later. In the meantime, nuclear confrontation was avoided, and both the American president and Soviet Premier Nikita Khrushchev agreed six months later on a nuclear test ban treaty. The latter was seen as a new symbol of U.S. and USSR revulsion for nuclear threats, as well as a possibility for nuclear disarmament in the near future. Consequently, the Cuban crises, from the Bay of Pigs to the nuclear threats of autumn 1962, represent the most threatening and dangerous moments of the Cold War. The Cold War toned down because of it, but the capitalist versus Communist ideological debate continued.

Note the student's command of fact and his or her ability to analyze the significance of fact. The relation between the Bay of Pigs invasion and the Cuban Missile Crisis is important, and it is adequately drawn here.

(5) Campaign Reform Act (1974): Influenced directly by public reaction to the Watergate scandal of 1972-74, the Campaign Reform Act represented the first law ever written to govern the use of money in presidential campaigns. It also established a Congressional Ethics Committee, set limits and guidelines for staff members in both the executive and legislative branches, and eliminated so-called special privileges held by members of Congress as well as executive branch cabinet officials. It was the most sweeping legislation of its kind ever passed, stressing ethical conduct for both campaigners and the victors of those campaigns. It was especially championed by a new generation of young liberal Democrats, such as Gary Hart and Pat Schroeder, who came to power promising great reform after the Watergate scandal and President Richard Nixon's August 1974 resignation.

Unfortunately, as some critics of the law suggest, such as 1992 and 1996 independent presidential candidate Ross Perot, the Campaign Reform Act also had many loopholes. The Political Action Committee (PAC), for instance, flourished after the law was passed, for the law made it difficult for members of Congress to accent precise issues, causes, and resulting "pork barrel legislation" without raising "special interest" or ethics charges.

Although far from perfect, the act addressed the issues of Big Money in presidential campaigns, as well as the abuse of power matters raised by the Watergate scandal. Unique and unprecedented, it represents the only legal legacy of the Watergate period.

Avoid the temptation to answer exam questions in terms of black and white or other absolutes. This answer shows a thorough appreciation of the very mixed blessings of the Campaign Reform Act; furthermore, it explains the pros as well as the cons rather than merely enumerating them.

Section B *(70 percent):*

Select ONE of the following. State your thesis clearly and be precise.

(1) To some post-World War II presidents, the Cold War required nothing less

than total American victory. To others, it meant finding a means of "coexistence" with communism. Why was anti-communism the backbone of U.S. foreign policy after World War II? Why was ideological confrontation with the communist nations so important to Washington?

(2) The presidency became more and more powerful during the post-World War II period, and especially during the Vietnam/Watergate era. Some say a powerful presidency offered efficient government and benefitted key policies, such as civil rights. Others say it led to corruption and national security abuses. What do you say?

I advise students to state their position early in the narrative, then argue chronologically in the name of logic and organization, embrace rather than avoid detail, and draw conclusions from the data. This step-by-step approach creates a persuasive piece, and that often means an excellent grade. Essay writing is much more personal than the multiple guess approach, and sometimes the announcement of some "Golden Rule" for successful analytical writing can do more harm than good. Trial and error might be the only Golden Rule here, although I try to avoid blatantly stating that fact in front of a struggling survey student.

ANSWER

(2) Shortly before the death of Franklin Roosevelt, the wartime White House and the Congress had reached a certain detente in the age-old battle of executive versus legislative control over leadership in policy-making. In court cases ranging from *Korematsu v. U.S.* to *Baumgartner v. U.S.*, the presidency had increased its "executive privilege" powers legally. World War II, of course, had constitutionally granted the commander-in-chief great command of the country' s political and economic direction. But whereas a Woodrow Wilson rejected Congressional input as "interference" during World War I and paid a sad political price because of it, FDR learned from Wilson's mistakes and welcomed Congressional participation in key war decisions if pressured to do so. His approval of the Fulbright-Connolly Resolution, for instance, near the end of his life, wedded him to full cooperation with Congress for the remaining months of the war. It demonstrated democratic harmony in the face of fascist tyranny, and, therefore, was good propaganda if nothing else. It would have been unlikely that Wilson would have (in his words) "submitted" to Congressional cooperation.

The next president, Harry Truman, given his confrontational style and also given the frustrations of Cold War jousting with the Soviets, saw presidential power increased with the Taft-Hartley Act and also saw majority public opinion and even key Congressional leaders, such as Republican Senator Arthur Vandenberg, pushing for more presidential leadership in the Cold War and not less. Eisenhower, although he favored a quieter personal style of leadership, did not reject these powers, and he proved the point in foreign policy, such as in the attempted assassination of Indonesia's Ahmed Sukarno, and in domestic policy too, such as in his decision to send troops to Little Rock, Arkansas, in defense of school desegregation.

JFK and LBJ, however, escalated the game, for they committed themselves publicly to the premise that their recent predecessors represented tired, do-nothing government. Promising everything from Cold War victory to man on the moon, JFK, for example, saw these ambitious policies always championed by a powerful president alone. LBJ shared a similar mission, and his Tonkin Gulf Resolution, which granted him special powers to conduct the Vietnam War, especially proved the point. Dismantling the resolution, especially after the war produced thousands of U.S. casualties, was difficult to do for Congress. By the time of the Nixon administration, many Congressional and public attacks on presidential power were interpreted by the White House as near-treasonous efforts, and in a time of war to boot. If he had not resigned in the Watergate scandal, Nixon' s defense lawyers had planned to use "executive privilege" as a justification for all alleged presidential wrong-doing during the scandal (for the U.S. had indeed been at war during the Watergate break-in and initial cover-up. Hence, the commander-in-chief enjoyed especially strong powers of privilege during war).

For all effective purposes, it could be said, depending on one's politics, that "executive privilege," FDR through Nixon, created more problems and confusion than necessary, thereby hurting efficient policy-making in both foreign and domestic affairs.

Since the students have been given other "executive privilege"-related questions in other exams and even in Section A of this very exam, many decide to continue the "tale" and answer this question as well. Indeed, the vast majority picked this question on a recent final exam in this course. Theoretically, their analytical skills should be at their best in this final, for it is their third exam within this format. A continuation of the "executive privilege" theme helps them organize their thoughts and data quite well if they put their minds to it. In the above answer, it is expected that the students also explain the mechanics of, for example, the Tonkin Gulf Resolution or the Little Rock crisis. This will demonstrate to the reader (me) that they know the detail and can truly

make a worthy conclusion from it. Especially in a politically charged question such as this one, some students exhibit strong opinions. Hence, some of them have difficulty moving beyond broad opinions and overly general comments. But they also know that a true analysis welcomes opinion if it is truly proven in the body of the essay. Consequently, there are plenty of students who know the difference and offer decent analyses.

The best survey students are the ones who learn from their essay-writing struggles of the previous exams and integrate both data and fine conclusions on the final exam.

CHAPTER 16

UNIVERSITY OF CONNECTICUT

HISTORY 231: AMERICAN HISTORY TO 1877

Wendy St. Jean, Lecturer

THE OBJECTIVES OF THIS COURSE ARE TO PROVIDE THE STUDENT WITH A GENERAL overview of some of the major developments in American history between 1492 and 1877. This course aims to heighten students' sense of the place of their own culture within a larger historical continuum. Many diverse cultures are compared in each segment of the course. In the first section, the major emphasis is on the conflict of cultures between Native Americans, African-Americans, and Europeans. In the second part, we look at the growing cultural divide between rural and urban Americans and between social classes. The last segment of the course considers the rise of cultural sectionalism between the North and the South and its manifestation in Civil War and Reconstruction.

Emphasis is on learning how to read both primary and secondary sources critically. Students should become sensitized to authors' biases and to writers' use of evidence to support their cases. I also want to familiarize students with key people and events that shaped early American history.

There is a midterm and final. Short answers and essays require students to assimilate what they've learned from primary and secondary readings to draw conclusions about the major issues considered in the course.

MIDTERM EXAM

Part I. *For fifty points (ten points each), answer the following:*

1. Discuss relations between the Powhatans and the Jamestown colonists. What key factors led to outbreaks of violence?

ANSWER

The Jamestown settlers were undisciplined soldiers who, instead of farming, tried to bully the Powhatans into feeding them. They committed outrages against the Powhatans, and the Powhatans may have interpreted the plague of 1616-18 as a conscious attempt on the part of the English to eliminate them. The expansion of tobacco cultivation was probably the key grievance that triggered the Powhatan War of 1622.

2. What military and strategic problems faced Great Britain in winning the American Revolution? Name at least five. What unorthodox tactics did American soldiers employ against the British army?

ANSWER

It was difficult and expensive to transport troops and supplies to North America; the colonists had only to defend themselves, not mount an offensive war; British troops antagonized the countryside when their supplies ran short; the British could not tell the difference between their friends and enemies; Hessian mercenaries were unreliable and often deserted to the Patriot side; North America lacked a great political center that the British could crush in a single blow; Spain and France formed an alliance against Great Britain; the British faced a concealed, armed enemy that employed unorthodox tactics like guerilla warfare, ambush patrols, targeting officers, sniping from behind trees.

3. How were Salem witchcraft outbreaks different from other witchcraft outbreaks that plagued New England in the seventeenth century?

ANSWER

(1) More people were accused: several hundred persons were accused in Salem (nearly 200 formally charged and imprisoned). (2) Some of the accused came from high social strata. (3) Accused people came from beyond Salem's borders (George Borroughs was in Maine at the time), and accused persons were unacquainted with the "possessed" girls.

4. Draw from Ethan Allen's life experiences to explain how radical the American Revolution was. Did Allen support Shays' Rebellion? Why or why not?

ANSWER

On the surface, Ethan Allen's experiences would suggest that the Revolution was a radical class war between social classes, i.e., landless farmers and rich speculators. However, Allen's exclusive concern for Vermont's future suggests that localism limited the Revolution's radicalism. Allen did not help Shays by sending men. He compromised—cut a deal—with New York authorities, so that they would make Vermont a state if the Green Mountain Boys promised not to join the Shayites.

5. Describe the British alliance with the Shawnees in 1783 and 1812-13.

ANSWER

The 1783 Treaty of Paris angered the Indians of the Northwest because it ceded their lands; however, the British maintained their alliance with the Shawnees and other groups by promising them military support to resist American expansion into the Ohio Valley. The British aided the Shawnees after the American Revolution because the Ohio Valley was useful to them as a buffer between the American nation and British Canada. During the War of 1812, the British armed the Shawnees until the British were forced to surrender in 1813. After that date they abandoned their Indian allies completely. British defeat in the War of 1812 ended Britain's presence in North America.

I tell my students to write practice outlines and essays. It's one thing to study the material, and another to be able to integrate it into a well-organized essay. I ask them to think about what issues I've dwelled on the most—religion, culture, politics, etc. I suggest that they group together lecture notes that touch upon similar issues and events. I will frequently ask them to compare two sets of leaders, conflicts, or cultural groups.

Part II. *For fifty points, answer one of the following:*

6.a. Explain how imperial rivalry and religious and commercial motives shaped the development of the Virginia and Massachusetts Bay colonies: What were the motives of the colonies' founders and the settlers? How did religion and commercial factors influence settlement patterns? How did these colonies' respective economic systems influence community development?

Here's how I think about grading this question. 10 points: Explain that Spain and Great Britain were contending for imperial power and that this rivalry encouraged British colonization of the New World. 10 points: Explain that the Virginia Company and settlers hoped to get rich quick by discovery of gold, a staple crop, or a passage to China. 10 points: Explain that Puritans came to American so that they could practice their religion the way they wanted and to set up moral communities. 10 points: Virginia's tobacco staple crop system resulted in dispersed settlements, little institutional development, class stratification, and widespread slavery. 10 points: New England's subsistence farming economy and religious goals resulted in denser communities with more religious and educational institutions, less class stratification, and slavery. Over time, New England communities emerged that were more commercially—and less religiously—motivated.

ANSWER

At the end of the sixteenth century, Spain had already conquered a vast empire in the New World that included South and Central America, Mexico, Florida, and the Caribbean. Locked in a power struggle with Spain, England initially sought to establish outposts or colonies in North America with a view toward raiding Spanish commerce. The first English settlements in the New World — Roanoke and Jamestown—were established as military forts rather than as agricultural settlements.

Think about when and how these categories (imperial, commercial, religious) apply. In what order should they be considered? In class lectures, in what order were they discussed? One of the first lectures focused on imperial rivalry, so you should probably know to discuss that first. Most of the lectures on Virginia focused on commerce/slavery, so those issues should be emphasized in relation to that colony; likewise, the lectures on New England stressed the role of religious ideas and moral principles in shaping the community.

Group your ideas into an outline, jotting down dates, ideas, and details as you think of them.

Reread the exam to make sure that each subquestion is covered in your outline and that you are not deviating from the question to include information that you do know that was not required. Go back to the outline and think about how to make transitions between paragraphs (ideas) and how to tie things together at the end.

Commercial motives provided a primary inspiration for planting the first English colony in North America. The early settlers of Jamestown (1607) hoped to find gold in the region, exploit Native American labor, and to reconnoiter a Northwest Passage to China. The vast majority of colonists were gentlemen or skilled craftsmen. When the Algonquian tribes would not feed them, the Jamestown colonists faced starvation. Nearly 70 percent died the first year of malnutrition and typhoid. Within the first decade of settlement, Virginians found a commercial commodity to market overseas—tobacco.

The expansion of this crop, so central to the colony's viability, resulted in hierarchical class and racial stratification. Most emigrants were either gentlemen or indentured servants, and planters began to import African slaves in large numbers toward the end of the seventeenth century. The headright system and necessity of farming along the rivers' edge resulted in a scattered, thin settlement pattern. Because of the dispersal of people, comparatively few Virginians attended (Anglican) church or school. This did not change until the Great Awakening the 1740s when evangelical churches offered an alternative to the church of the planter and royal establishment.

Religious motivation proved central to the Puritans' "Great Migration" to Massachusetts from 1630 to 1642. Puritans feared that their children were in danger of corruption by English churches that had not carried the Reformation far enough. The New England settlers' sense of religious mission contrasted with the dominant economic motives of most Chesapeake settlers. Puritan religious beliefs left their mark on New England settlement patterns. The majority of New England colonists came from the middling classes and included a broad range of ages and a fairly equal gender ratio. Puritan emigrants set up subsistence agricultural communities around meeting houses. The Massachusetts Covenant and the colony's blue laws reflected their determination to forge a model of moral order. They relied primarily upon the family, rather than upon servants or slaves, as a labor force.

In the second half of the seventeenth century, wealthy commercial communities developed in the seaports of Boston, New Haven, Gloucester, Marblehead, and Salem. Around 1690, speculators in New England, driven more by economic motives than religious ones, began to alter settlement patterns in favor of more scattered communities.

On balance, while the colonization of British America undeniably occurred within a larger framework of imperial rivalry, commercial and religious motivations predominated in the shaping of New England. In Virginia, commercial motives were most central to the enterprise. African slaves, who made up about 20 percent of inhabitants of Virginia in 1700, were part of a commercial, and not a religious, operation.

This student answered each part of the question; that is, she looked at imperial rivalry, religious motives, and commercial motives. She compared these factors as they applied to the Massachusetts Bay and Virginia colonies and provided dates of crucial events. Note that she has

pulled in evidence from readings (Bernard Bailyn's *Robert Keayne*, Karen Kupperman's *Starving Time at Jamestown*, Virginia DeJohn Anderson's Great Migration). Although she drew distinctions between the two colonies, she also recognized that neither colony was static. Over time, the Massachusetts Bay grew more commercially oriented and Virginia more religious.

6.b. Argue that slavery was integral to early American development from the 1670s to the 1770s: How did American slavery fit into the worldwide commercial system? Why did planters prefer African slaves to white servants or Indian slaves? How did slavery in the Middle Colonies and New England compare? Was slavery widely accepted?

For 10 points: Slavery not unique to North America, but part of a world-wide system of unfree labor. 20 points: Explain transition to slavery—whites hard to come by because conditions improved in England and other colonies offered better deals; Africans brought agricultural skills and had greater endurance and immunity to malaria; Indians ran away, and there was a threat of warfare from Indian tribes. 5 points: Mention Bacon's Rebellion (1676) as a turning point that alerted planters to the danger of system of white servitude, and the Yamasee War (1715) as signaling an end of Indian slavery. 10 points: Describe prevalence of slavery in Middle Colonies and New England and also mention general attitude toward slavery. 5 points: Explain Enlightenment critique of slavery and its relatively minor impact on course of events.

ANSWER

American slavery was part of a worldwide system of unfree labor, which included slavery, indentured service, and serfdom. In the fifteenth century, the Portugese introduced "new Slavery" in the Atlantic sugar colonies—vast numbers of African slaves indentured for life and employed in large-scale commercial agriculture linked to the Atlantic trading system. The overwhelming majority of the some 10 to 12 million slaves shipped to the New World from 1500s to the 1880s worked on sugar plantations in Brazil and the Caribbean. Although a far lesser number labored in North America, their presence proved crucial to the development of the thirteen colonies.

First: Think about when and where slavery emerged as a crucial issue in lectures and discussions. Think about the time frame and what events transpired at ends and in the middle of these dates. Second: Write outline. Third: Review questions and make sure you've included each subsection of the question in turn. Fourth: Ask yourself how the separate questions all relate. Make sure you are not only answering each subsection, but also keeping the broader question in mind. Make sure you don't go off on a tangent about one issue and forget that the question

should aim toward a particular end. Use the overall question to tie together paragraphs and bring back separate ideas to the larger point.

In the 1670s, Virginia planters turned increasingly from the indentured servitude of whites to the importation of African slaves. South Carolina, founded in 1670, began using African slaves that many settlers brought with them from their plantations in Barbados. Indentured servants proved unruly and few were attracted to the Chesapeake or South Carolina in the late seventeenth century because economic conditions in England had improved, and New England, New York, and Pennsylvania offered enticing alternatives. African slaves proved more resistant to malaria and the hot, humid weather of the Southeast. Their knowledge of rice and indigo cultivation, animal husbandry, hunting, and fishing was superior to that possessed by white immigrants. African slaves helped lessen the class tensions that existed in Virginia between landless former indentured servants and wealthy planters. Bacon's Rebellion of 1676 alerted planters to the need for a new labor source. American Indians knew the terrain and therefore could escape more easily than Africans. The enslavement of Indians also caused tensions between English settlers and Indian groups that threatened and sometimes resulted in warfare. After the Yamasee War of 1715, the practice of enslaving Indians died out. In the southern colonies, profits from African-produced staple crops such as tobacco, rice, indigo, and sugar fueled economic development.

Slavery also existed in the Middle Colonies and New England in the colonial era, although in smaller numbers. Thirty thousand slaves inhabited Pennsylvania, New York, and New Jersey by 1770, and five thousand lived in New England. Northern religious sensibilities were not offended by the institution. Puritans owned slaves, regarding their lowly station as ordained by God.

Aware of slavery's Biblical precedents and economic importance, white settlers North and South, accepted slavery throughout most of the colonial years. An Enlightenment critique of slavery did emerge in the eighteenth century; however, the *philosophes*' most avid readers—Thomas Jefferson, James Madison, and other Virginians—were prominent slave holders. Because slavery was so crucial to the colonial economy, the antislavery ideas had a limited influence in America. When American Patriots protested Great Britain's conspiracies to deprive them of their liberties, they often likened themselves to slaves. American Whigs did not necessarily condemn chattel slavery of blacks, but rather the Crown's attempt to reduce free Englishmen like themselves to such a degraded status.

American thinkers had a great fear of dependency, and in the view of some political thinkers, slavery ensured American freedom. The enslavement of blacks left white men free of control by other white men.

Slaves were influenced by revolutionary ideology and fought on both sides of the Revolution in quest of their freedom.

The student addressed each subsection of the question, beginning in the 1670s and ending in the 1770s, just as the question asks. It is important to answer the question in the terms that the question poses. The student also explains why the dates specified are important markers. Her argument is backed with solid evidence and explanation. She's drawn upon and integrated assigned reading (Peter Wood's *Black Majority*, Edmund Morgan's *Slavery and Freedom: An American Paradox*, Patricia Tracy's *Ben Franklin's New World*).

FINAL EXAM

What will help students complete the course successfully? Take notes on the readings, attend regularly, participate actively in class discussion, and meet in study groups. Since I give them questions to consider each week, they should focus on these. If they do not fully comprehend the answer, they can raise questions in class or during my office hours. I am also very accessible through e-mail.

The midterm and final examinations account for half the course grade, the midterm 25, and the final 25 percent.

Answer all questions.

I. *For ten points each: Provide dates and specific details wherever you can.*

For the final exam, students have 2 hours. I recommend that they spend 40 percent of the time, about 50 minutes, on the essay and 60 percent on short answers—about 10 minutes on each. If they finish early, I encourage them to look over their answers again and to write in the margins of their booklets any information they forgot to include.

1. **Explain the shortcomings of the Freedman's Bureau from the perspective of African-Americans. How did President Johnson frustrate the Bureau's successful aid to former slaves?**

ANSWER

The Freedman's Bureau had to retract its promise to redistribute lands. President Andrew Johnson instituted a new policy of "restoring" lands to their former owners and revoked General William Tecumseh Sherman's Special Field

Order No. 15 (January 1865) and O. O. Howard's Circular No. 13 (July 1865) that provided for the redistribution of confiscated Southern plantations into 40-acre plots. The Bureau enforced black codes in some states that required blacks to return to work on their former master's plantation. The Bureau was underfunded, understaffed, sometimes corrupt, and usually lacked sufficient military backing to enforce fair labor practices. Sometimes Bureau officials shared Southern whites' belief that blacks should be restricted to hard or menial labor.

This is a typical short-answer question. A successful answer might involve the following steps:

1. **Think back on reading and class discussion; rephrase the question in your mind. This particular question tests whether you have read and taken notes from an assigned article, Stephen Nissenbaum's "An Insurrection that Never Happened: The Christmas Riots of 1865," *True Stories from the American Past* (Altina Waller and William Graebner, eds., McGraw-Hill, 1997, 1:254-70). Most credit would be given for making the central point that the Bureau did not give ex-slaves their own lands because the president ordered the Freedman's Bureau to reverse its former policy.**
2. **Ask yourself whether you have provided relevant dates and details. For full credit, you would need to provide a detailed answer with specific information (e.g., Sherman's order and Howard's circular).**
3. **Is there anything you can add to your answer? Do any less important points come to mind? Additional credit is given for inclusion of secondary issues such as corruption and financial problems that plagued the bureau.**

2. **Why did Lincoln announce the Emancipation Proclamation when he did? Why not sooner or later than he did? What did it accomplish? Give the date.**

ANSWER

Lincoln issued the Emancipation Proclamation on January 1, 1863. It freed the slaves in the areas of rebellion, but exempted slaves in the border states and former Confederate states under Union control.

Lincoln did not issue it earlier because he did not want to alienate the border slave states. (Lincoln was famous for saying he'd like to have God on his side, but he had to have Kentucky.) Lincoln also was waiting for a major Union victory (Antietam) before issuing the Proclamation; otherwise it would seem an act of desperation. There was also the momentum of total war and the initiatives of Northern generals, which accomplished de facto what the Proclamation stated. So Lincoln could not afford to wait much longer.

Lincoln needed to recruit black troops for the war effort. The Proclamation opened the floodgates of black enlistment in the Union Army. It transformed the war into a war for freedom. Given the antislavery sentiment in England, Lincoln also issued the proclamation to dissuade Great Britain from recognizing and granting military aid to the Confederacy. Now there was no chance that that country would ally with the South.

3. Why was the Election of 1864 a decisive event?

ANSWER

McClellan ran against Lincoln on a compromise platform. The Election of 1864 resulted in Lincoln's reelection. This election was interpreted by the president as mandate for his policy of "unconditional surrender." Lincoln's reelection demoralized the South, for few believed the South could hold out for four more years of fighting.

4. Why did Reconstruction fail?

ANSWER

The primary reason it failed was because blacks did not get their own land and remained dependent on whites. Mercantile interests North and South conspired against black property ownership and economic progress because they needed blacks to cultivate cotton. A secondary reason it failed was because of Northern and Southern racism. Northerners and westerners discouraged blacks from moving to their states. Even many former abolitionists no longer cared about blacks once they had gained their freedom; in other words, they opposed slavery but did not advocate equality for blacks. Another reason for Reconstruction's failure was class conflict in the North seemed more pressing to northerners than racial conflict in the South.

5. Why did blacks enlist? Why were they disillusioned? How did black soldiers aid the Union war effort?

ANSWER

Many blacks enlisted in the Union army after Lincoln issued the Emancipation Proclamation. These were some motives for enlisting: belief in the cause of freedom; wanted to prove manhood; retaliation against former masters; wanted to gain civil rights; economic need; pressure from their communities. Black soldiers were disillusioned because they were not treated as equals in the Union army. The large majority of officers were white, blacks received inferior guns and provisions, less pay, and were relegated to fatigue duty and other supportive roles. They were far more likely to die of disease than of wounds in battle. Black soldiers crippled the Southern war effort by leaving plantations for the army, thereby causing a labor shortage. Black regiments, reminiscent of slave revolts, frightened and demoralized the Southern communities. They contributed directly to Northern victory by their strength of numbers.

6. How did Northern and Southern cultures differ, and how were these differences rooted in their respective economic systems? What were the contrasting values of the two sections?

ANSWER

Northern and Southern cultures differed in these ways: (1) economic—North was more industrialized; South more agrarian; Northern economy based on free labor; the Southern economy rested on slave labor. (2) Politics—the North was predominantly Republican, South Democratic. (3) Ethnic—North had larger immigrant population, South had more ethnic (although not racial) homogeneity. (4) Cultural and Religious—North, greater literacy, more active reform movement; Southern religion less activist; also South had a "folk" society characterized by more oral traditions.

The contrasting values of the two sections: South emphasized martial values, paternalism, and honor. Northern ideals were influenced by the penetration of capitalism—men and women were to exercise restraint, hard work, and responsibility.

II. *For forty points, answer one of the following:*

7.a. Why is the theory of "internal dissent" an unsatisfactory explanation for why the South lost the Civil War? What are other possible explanations (just list them)? How was that theory challenged by events in the North and the border states? Explain divisions in two Union-allied states.

I evaluate this answer in the following terms. 5 points: Some explanations include superior resources and numbers, leadership gap, loss of will. 10 points: Explain that "internal dissent" theory was the theory that the South was divided by class tensions and conflict over states' rights. 5 points: These divisions did not prove as troublesome to the Southern war effort as social divisions and racial conflicts were in the North. 10 points: Race, class, and political divisions in New York. 10 points: Ethnic, geographic, and class differences reflected in pro-Confederate versus pro-Union loyalty, divisions in Missouri.

ANSWER

Traditional explanations for Southern defeat: (1) Greater numbers and resources in North, (2) inferior leadership in South, (3) loss of will to fight/sacrifice too great, (4) internal dissent: South was divided into social classes and by the issue of states' rights.

Begin by asking why these seemingly unrelated questions are brought together—the South's defeat and Northern social divisions. Next, try to decide where the emphasis is placed—which part of the question seems most crucial. Then answer each subsection, but provide the most details where they are required—that is, on the central point of examining internal conflict in the *North*, not the *South*.

The last explanation was disproved by Northern victory. The North had even greater internal dissent, which surfaced in guerrilla warfare in border states and during the New York Draft Riots (1863). Draft Riots exposed class, ethnic, and racial conflict in New York City. Irish-Catholic workers hated blacks for taking their jobs and hated the rich for cutting their wages. Rioters lynched blacks and burned essential war-production factories. Irish workers were Democratic voters and opposed the efforts of Republican bosses to reform them by advocating temperance and other moralistic legislation. The draft was particularly offensive because of a commutation clause by which the rich could pay a fee to waive their service. In Missouri, free-labor German and Yankee farmers of the North battled pro-Confederate planter-dominated guerrilla forces in the Southwest. New York and Missouri are just two examples that illustrate the lack of unified support behind the war effort in the Union states.

A good answer successfully addresses each part of the question. The easy part is listing the possible causes of Southern defeat. The real challenge is to understand the social divisions in the free states. Some students will not make the connection that Northern examples of internal conflict disprove the argument that internal dissent weakened the South. Most students will remember that New York was torn by the Draft Riots, but fewer students will remember that Missouri was a state that remained with the Union. The real challenge, however, is describing in detail the bases of the social divisions in these states. This student has done a good job of integrating course readings into his answer.

7.b. Explain the economic/social origins of the Strike of 1877: What were the grievances of the workers on the B&O line? What did strikers demand from the owners of the B&O? What values did strikers and their supporters feel

they were defending? What values did middle class businessmen and their allies believe they themselves were defending? What was President R. B. Hayes's role in the affair? Were strikers' gains or losses greater?

10 points: List grievances and demands of railroad workers. 10 points: Explain that workers opposed monopolies and sought promise of free-labor ideology—social mobility; also drew upon rights ideology to seek a living wage. 10 points: Explain the perspective of businessmen and the middle class, their low esteem for working class, especially immigrants; their view of free-labor ideology was to put the blame on the victim—the worker was starving because he wasn't working hard enough or was drinking away his wages. 10 points: Outcome—use of federal troops to put down strikes, construction of armories, publicity for and against workers, increase in size of unions.

ANSWER

The economic and social origins of the Strike of 1877 grew out of the Civil War. Factory owners profited from rapid economic growth, while the conditions of workers declined. At end of the war, owners enjoyed excess of labor and cut wages below subsistence. The "Great Upheaval" began in West Virginia when the B&O railroad managers announced a 10 percent wage cut; they had already cut wages eight months earlier. Workers also wanted to abolish the classification system and the practice of assigning trains. They sought stabilized hours and pay. The company's stock prices were rising, indicating that it was trying to squeeze its workers for all the owners could get from them. Strikers were trying to gain better circumstances for their families and wanted to fulfill their aspirations of social mobility. They drew upon republican, free-labor ideology to argue that economic and political inequality was too large between workers and owners. They protested the rise of monopolies. Strikers and their supporters defended their right to a living wage. The middle class blamed the workers' poverty on their own faults such as lack of restraint (working immigrant families were larger), alcoholism, laziness, innate deficiencies based on their ethnicity/race. Businessmen claimed the constitutional sanctity of property.

1. **Decide which of the subsections is at the heart of the essay and spend the most time thinking about that one.**
2. **Answer each part of the question; failure to answer each part will result in loss of points. Where you don't know the exact answer, do your best to write something, anything. You may get partial credit if you at least take a stab at it.**
3. **Double check to make sure you've addressed each subquestion as well as the larger**

overall issue.

4. If you have time remaining, consider what other details from your reading (Belcher, *Great Upheaval*; film, *1877: Grand Army of Starvation*) you can add.

President Hayes had no sympathy for the strikers. He used federal troops to crush the strike. This set a precedent for government intervention on behalf of business in labor disputes. Thenceforth, major cities would also have armories to protect businessmen's property from the "mob." Workers lost because government lined up with business, and they were stigmatized with the blame of having caused riots. At the same time, workers successfully publicized their grievances, and unions expanded.

This answer works well because it illuminates the different perspectives of workers and the middle class on the Strike of 1877.

CHAPTER 17

DUKE UNIVERSITY

HST 91D OR 92D: DEVELOPMENT OF AMERICAN DEMOCRACY TO 1865

Kathryn Fenn, doctoral candidate

MIDTERM EXAM

THIS EXAM IS FOR A COURSE COVERING THE COLONIAL PERIOD UP TO THE CIVIL WAR. The main purpose of the course is to introduce students to the major themes in American history and to teach them the methods of historical writing, research, analysis, and argumentation. Goals of the course include developing ability—

1. To recognize the major periods, events and themes in American history and to learn how to connect individual historical events or concepts with their historical context and with the larger themes of American history.

2. To learn how to ask key historical questions (such as who, what, when, where, why, and how) and how to construct a historical argument and a historical paper or essay. A key aspect of this is understanding the differences between primary and secondary sources, and to learn how to use them in constructing a historical paper.

3. To learn how to use the resources at a library to undertake historical research.

4. To recognize that American history is made up of a number of specialized subdisciplines (such as women's history, African-American history, Native

American history, military history, environmental history, economic history, cultural history) and to introduce students to the differing emphases of and questions raised by the practitioners of these subdisciplines.

Please take a few minutes to check over your work and to proofread before turning in your exam. Good Luck!

Heed the instruction to proofread! Plan your time so that you have a few minutes to go over your work in order to catch obvious grammatical errors and to ensure that you have not left out important words. Hasty errors can undo even the best work.

Part One: *Identifications (take 15 minutes, worth 30 points total)*
Identify, in several complete sentences, FOUR of the following terms. Be specific about the time and place associated with each term, clearly explain what the term is, and explain the significance of each for American history.

A. Gullah
B. John Locke
C. *Cohens v. Virginia*
D. Bacon's Rebellion
E. Anaconda Plan
F. Gag Rule

ANSWERS

A. Gullah is a language that is a mixture of a variety of western African languages and English. This combination or creole language emerged during the early to mid-nineteenth century, and was spoken by slaves in specific regions of the South where there were large concentrations of slaves, and many who continued to arrive from Africa—namely, the Sea Islands of South Carolina and Georgia. Its significance is that it represents the slaves' application of their West African cultural background to the new circumstances and culture in the New World to create a new, uniquely African-American culture. This process of creolization occurred not only in the realm of language, but also in the realms of cooking, building, clothing, and music.

B. John Locke was an eighteenth-century English philosopher whose ideas are associated with the theory of Natural Rights expressed most famously in the Preamble of the Declaration of Independence. The theory of Natural Rights is embodied in the phrases "All Men are Created Equal" and "are entitled to Life, Liberty and the Pursuit of happiness" in the Declaration of Independence. Locke's writings were well known by the men and women who struggled to lead the American Revolution, and later to frame the government of the fledgling nation. Many parts of the Constitution that are based on the theory of Natural Rights proved later to be quite troublesome—as would quickly become the case with the issue of slavery.

C. *Cohens v. Virginia* was a Supreme Court case that was decided in 1821. In this case, Cohens was found guilty by the Virginia state court of illegally selling lottery tickets, and he appealed the case to the Supreme Court. The Court upheld Virginia's conviction, but the case established the right of the Court to review and reverse the decision of state courts involving the powers reserved to the federal government, such as the regulation of commerce. This precedent would lead to much legal and political turmoil regarding the issue of states' rights in the future (particularly in the decades leading up to the Civil War, and beyond). Still, the principle of Federal Review still stands to this day.

F. The Gag Rule was a law passed by Congress in 1836, which declared that all petitions related to slavery and its abolition would be "laid on the table" and not discussed. This was a measure put through by Southern politicians and their allies to stifle discussion of the slavery issue, but it ironically increased the furor over slavery by alarming the abolitionists into new levels of action, since they saw the Gag Rule as proof that the "Slave Power" had taken over the federal government and was using it to further its goals. The Gag Rule thus marks an across-the-board intensification of pro- and anti-slavery activity and was one of several reasons why the conflict over slavery became irreconcilable through traditional political means. Other reasons included economic changes, which made slavery more profitable, and the Evangelical revival in the North.

What makes these good answers is that they clearly identify what the term is about, including information on who, when, and where. Also, each answer examines the larger significance of the terms, and raises other issues and concepts from the course in order to enlarge on the particular term's significance, and to tie together the course by focusing on the issues that arise again and

again, in different times and places. Furthermore, the answers are the right length and have the right balance of information: two identifying sentences and two sentences on significance.

Part Two: *Essay (take 35 minutes, worth 70 points)*

Choose one of the topics listed below and write a concise, well-organized essay on it.

A. In the 1820s and '30s, there were two convergent developments: the growth of cities and industrial production, and the rise of a variety of reform movements (including temperance, abolitionism, labor activism, dietary reform, "domestic feminism," etc.). Choose a minimum of three of these (and/or other) reform movements, describe their goals and activities, and then develop an argument about how they were related to specific aspects of urbanism and industrialism.

B. In this course we read many accounts from the nineteenth century about what life as a slave was like. Some of these accounts were written by slaves or ex-slaves, and some by non-slaves. Discuss some of the themes and the information in these accounts as it pertains to the slaves' desires and attempts to gain freedom. Develop an argument on the topic of freedom or escape from slavery, and be sure to place this argument in the context of the sources and the time period.

The skills involved in historical argumentation, research and in constructing class papers are directly called upon in the essay questions, which appear on both the midterm and final.

ANSWERS

B. Slavery and Freedom: The Means to an End

Many primary sources by and about African-American slaves in America describe or discuss the slaves' strong desires to be free. But these accounts also contain many stories about slaves actually taking the steps to become free. Some of the accounts provide detailed information about the means by which slaves emigrated or escaped to freedom—means developed and disseminated collectively over many decades. Although it could not guarantee successful emigration, a body of knowledge about both the possibility and means of freedom did exist among slaves, and seems to have penetrated every corner of the slaveholding South by the time of the coming of the Civil War. The existence of this body of knowledge should be factored into our understanding of the culture of the antebellum South, as well as the nature of the conflict over slavery that led to the Civil War.

The first paragraph defines the content and significance of slave narratives. In so doing, it demonstrates not only the student's appreciation of the role these narratives play in documenting slavery—and in disseminating useful information on escaping from slavery—but also an understanding of the function of primary sources in the study of history.

The emigration of African-American slaves to freedom was in many ways an extension of a mobility within the South that some slaves had been able to establish over many generations, despite tremendous constraints. According to historian Douglas Egerton, the ability of slaves to hire themselves out, and thus move about both the city and countryside with relative freedom, was a well-established feature of slavery during the eighteenth and into the nineteenth centuries. Furthermore, Egerton describes this "twilight world between slavery and freedom" as the breeding ground for schemes of rebellion and escape.

John Blassingame's collection entitled *Slave Testimony* includes a few cases in which slaves were able to achieve freedom and then emigrate to free lands by buying themselves from their masters. In the majority of cases, however, this option proved impossible either because of cheating owners or any other number of reasons. Thus emigration meant escape for the majority of African-American slaves.

Before dealing with the dramatic subject of escape, the student points out that other means of "emigration" were sometimes available. Thus he builds to his strongest points rather than beginning with them. This gives the essay more impact.

There seem to have been two main methods of slave emigration, or escape to free lands: to travel through mainstream Southern, and sometimes Northern, society (which required disguise or dissimulation) or to travel outside of mainstream society (which required hiding and careful navigation from one safe spot to another). Of course, in any one slave's escape, both methods would most likely have had to be used. Still, I believe it is important to be aware of this distinction because different skills and knowledge would have had to be mobilized in each context.

This paragraph defines two methods of slave emigration, setting up much of the rest of the essay, which elaborates on these methods. This is solid organization: set out and define terms and topics, then elaborate on them with examples, statements of significance, meaningful exceptions, and so on.

In the case of the first sort of emigration (moving within mainstream society itself), successful emigrants would have needed a knowledge of the "free world" which gave them the skills and confidence to venture out in it. Thomas Likers, a waiter from Baltimore who "always [had] the privilege of hiring my own time," described his emigration in this way: "I didn't run away, but rode off in the cars. I paid my fare like a gentleman, and had no trouble in getting away." What is unspoken here is that Likers must have had a great store of knowledge to be able to both look and act "like a gentleman" in a way convincing to whomever sold him his tickets—knowledge he had the opportunity to gain because of his mobility as a skilled self-hiring slave and his contact with restaurant patrons.

In the case of the second sort of emigration practiced by slaves (moving outside of mainstream society), it was essential to have knowledge of the "secret world" of support available to fugitives. A good deal of this clandestine support was offered by institutions and their members, who provided shelter, transportation, and information to fugitives and would-be fugitives. The most obvious institution to mention in this respect is the Underground Railroad. It was described in 1894 by the *Detroit News and Tribune* as made up of "active agents," who went among the slaves in the South and encouraged them to run away to Canada. Mentions of the Underground Railroad or abolitionists generally appear scattered throughout Blassingame's accounts as initiators or helpers in attempts at escape.

One paragraph is devoted to each of the forms of emigration. These separate paragraphs keep the discussion focused and clear.

Harriet Tubman's story is a virtual chronicle of the skills and knowledge accumulated by those who were able to escape and help others escape. She learned a variety of practical skills for concealment, including traveling in the winter, adopting remote rendezvous points, navigating by the rivers and telling the time by the stars. She also learned to spy upon and foil the efforts of slave catchers, often through the use of her agents. With regard to her use of others, her interviewer noted that "she had confidential friends all along the road." In terms of traveling within society, Tubman knew how to look very respectable, not at all like a poor fugitive. She was also described as having "command over her face" and knew how to "banish all expression from her features, and look so stupid that nobody would suspect her of knowing enough to be dangerous." Finally, Tubman knew how to use guns and to fistfight, and was willing to use both a gun and her fists in defense of herself and others.

The consequences of this widely circulating body of knowledge about moving clandestinely within or outside of society pertain to both the ante- and postbellum periods. During slavery there were indeed secret worlds, institutions, and skills that nearly all slaves were familiar with, whether or not they utilized them. This attests to the resourcefulness of slaves who, along with their free allies, forged spaces in which they could exercise choices with regard to their future. Furthermore, the institutions, networks, and skills developed under the pressure of slavery would continue to play an important role after emancipation, when new constraints were thrust upon the African-American community. Any consideration of African-American social, cultural, or political activity in the postbellum era must take into consideration the unique skills and styles developed by slaves in their struggles to emigrate to freedom.

What makes this a successful response is, first and foremost, that it has a clear topic and argument in the first paragraph that is referred back to throughout the essay, and a well-organized conclusion that responds to the argument. This answer is successful also because all of the major points are supported by specific sources and historical examples. Additionally, this response refers to some of the major arguments in the historiography on slavery—which makes it an excellent, or A-quality answer. Finally, the essay is of adequate length, and it has a clear, appropriate title. When 35 minutes is given for an essay, I expect there to be a minimum of three pages of writing—that is, three "blue book" pages. This would also equal five to nine paragraphs.

FINAL EXAM

This exam is for a course covering the post-Civil War period up to the present.

In terms of percentages, the final exam and midterm together count for over half the class grade (55 percent), and the exams are important as moments when the degree to which students have been doing the reading and thinking about class discussions is brought to light. Improvement from the midterm to the final is also a desirable thing. This often factors into the

final grade in terms of overall class participation or in rounding up a grade that is on the border between two grades.

Please take a few minutes to check over your work and to proofread before turning in your exam. Good Luck!

Part One: Identifications *(take 35 minutes, worth 25 points total)*
Identify, in several complete sentences, FIVE of the following terms. Be specific about the time and place associated with each term, clearly explain what the term is, and explain the significance of each for American history.

A. Scopes Monkey Trial
B. Levittown
C. Anne Moody
D. Port Huron Statement
E. NSC-68
F. Stagflation
G. "Kitchen debate"
H. Huey Long

Success on examinations depends on preparation and active participation throughout the semester. Attend class regularly and be prepared to contribute to class discussions. Attendance and participation are extremely important, both because they are the basis for getting the information students need in order to do well and also because they show the instructors that a student is interested and involved in the class.

Also, class papers are very important as demonstrations of how well students understand the major themes and concepts and how closely they're doing the readings. Planning ahead on the papers so that you have plenty of time to do the research and recover from unexpected difficulties in doing the research is central to success. Students who can get me a draft of their term paper at least a week before it's due, so I can give feedback, will always make an improvement in their grades.

ANSWERS

A. The Scopes Monkey Trial occurred in Tennessee in 1925 and addressed the issue of whether or not evolution could be taught in classes. John Scopes, a teacher in Tennessee, broke an anti-evolution law by teaching Darwinism. He was defended by Clarence Darrow of the ACLU, while the prosecution was led by the great orator and Fundamentalist William Jennings Bryan. The hearings were aired on the radio and so received unprecedented attention and raised a nationwide debate. Even though Scopes's conviction was upheld, the evolutionists greatly aided their cause by making Bryan look foolish. Still, the trial highlighted the growing polarization not only of creationists and evolutionists, but rural folk from urban folk, such that the rural creationists were strongly opposed to the teaching of evolution while the urban evolutionists felt the teaching of evolution was essential to having a well-educated population and promoting scientific progress.

Though many Americans were persuaded by the evolutionists' argument, the creationists fought to hold onto a belief system and a way of life that they saw threatened by urbanization and other forces.

B. Levittown was the first mass-produced suburban neighborhood, built in the 1950s on Long Island, New York, by Alfred and William Levitt.

Levittown's nearly identical houses arranged in perfectly symmetrical rows along straight streets suggested a mindset of conformity. At the same time, though, the space provided by the affordable Levittown houses and yards became a symbol of Americans' postwar affluence and also indicated the need for increased housing due to the baby boom. These postwar suburbs, though an extension of the earlier "trolley town" suburbs, were able to develop on a much larger scale in this period because of the advent of the highway system and the increasing ability of Americans to afford automobiles. The massive postwar flight from urban centers furthered the geographic segregation of whites and blacks and led to increasing racial tensions as the quality of life in the cities deteriorated.

D. The Port Huron Statement was issued in 1962 by the Students for a Democratic Society. Within the statement, the students explained their beliefs, including the idea that the government had strayed too far from the people it represented. Further, it called for an end to racial segregation and discrimination and the Cold War. The statement also discussed ways these goals could be pursued within the existing social and governmental systems.

In contrast, the Manifesto written by members of the Weather Underground in 1969 was much more radical—calling for the overthrow of the U.S. systems of capitalism and imperialism. Thus, the Port Huron Statement serves as an example of the moderate protest and reform ideas of middle class college students in the early '60s.

F. Stagflation is a term used to describe an economic situation that developed in the 1970s in which high inflation was combined with high rates of unemployment. Normally these two economic indicators move in opposite directions from one another, but a unique combination of events in the '70s led to this unusually bad economic situation.

Firstly, the U.S. was undergoing a massive transformation from an industrial-based to a service-based economy. Secondly, U.S. firms were for the first time since World War II facing stiff competition from their European and Japanese counterparts. Thirdly, investment in the U.S. was sluggish. Finally, and perhaps most significantly, the decision of the OPEC cartel to embargo the sale of oil to the U.S. as a punishment for the U.S. support of Israel in the Yom Kippur War made energy prices skyrocket and further slowed the economy. The resulting stagflation is significant because it was the first crack in America's global economic dominance since World War II and it helped to usher in a more conservative and somber mood in the '70s, contrasting to the mood of optimism and reform of the '60s.

H. Huey Long was both a senator and governor of Louisiana who, during the 1930s, created and implemented a plan for alleviating the hardships of the Depression called "Share Our Wealth." Long's plan capped any person's annual income at $1 million, with any excess collected by the government and redistributed to the poor so that everyone would be guaranteed a minimum annual income of $2,500. Long also advocated a government-provided free college education for all Americans and a 30-hour work week. His plan gained support in the 1930s and he might have posed a challenge to FDR in the 1936 election had he not been assassinated in 1935. Long's "Share Our Wealth" plan provided an alternative to the New Deal, and provides a picture of a road not taken during this period. Likewise, the widespread support for his unorthodox plan illustrates the depth of the desperation caused by the Depression and the willingness of some Americans to accept a radical redistribution of income as a way of solving social and economic difficulties.

These identifications are successful for the same reasons given in the comments on the midterm for this course; however, note that these identifications are somewhat longer than those in the midterm—as should be expected, because more time per answer is allotted in the final. Identifications in the course final should be five to eight sentences long, should go into more detail in identifying the term, and should elaborate at more length on the significance of the term or concept in question.

Part Two: Quote Explication *(take 25 minutes, worth 25 points)*
Choose one of the following quotes and explicate its meaning.

A. "I do not like subversion or disloyalty in any form, and if I had ever seen any, I would have considered it my duty to have reported it to the proper authorities. But to hurt innocent people whom I knew many years ago in order to save myself is, to me, inhuman and indecent and dishonorable. I cannot and will not cut my conscience to fit this year's fashions, even though I long ago came to the conclusion that I was not a political person and could have no comfortable place in any political group." —*Lillian Hellman, 1952*

B. "The feminists want to abolish the family. But the family is the basis of everything. It is the foundation of our society; if that crumbles, everything else goes." —*a pro-family activist, interviewed by Rebecca Klatch in the 1970s or '80s*

ANSWERS

A. This is a quote from the playwright Lillian Hellman from her testimony in front of the House Un-American Activities Committee in 1952.

Hellman was asked to testify about her own political beliefs and those of her friends before HUAC during the post-World War II Red Scare. During this period it was typical for artists to be investigated for subversion. One example of this is the Hollywood Ten, a group of actors and directors who served time in jail for refusing to testify before HUAC.

In this quote, Hellman voices her objections to the tactics of HUAC, claiming that their attempt to force people into "turning in" their friends and acquaintances in order to save themselves is an "inhuman, indecent and dishonorable" thing. She argues that HUAC's anti-Communist searchings are "this year's fashion"—a turn in the political tide—and that she, by contrast, is upholding humanity, decency, and honor by not betraying her innocent friends. Hellman thus turns the tables on the investigators' logic, presenting herself as a more traditional American, and, in doing so, helped to foster a shift in public sentiment against the red-baiting tactics of republican Senator Joseph McCarthy and others in HUAC.

Margaret Chase Smith is another figure who denounced HUAC's bullying tactics and "trial by accusation."

This quote provides insight into both the investigations of Communist subversion in the early '50s and the negative response to it by members of the artistic community. Hellman's courageous and clever position helped to sway the opinions of many politicians and the public at large, to see the leaders of the Red Scare as the truly "Un-American" ones. This led to the public turning against McCarthyism during the 1954 Army hearings into the HUAC trials.

This quote explication is successful because it clearly identifies the source and context of the quote, addresses the content of the quote itself—that is, refers to what is being said in the quote and explains it—and considers the larger significance of the subject matter. Also, three paragraphs is about the appropriate length for this type of answer.

Part Three: Essays *(take 60 minutes, worth 50 points)*
Choose one of the topics listed below and write a concise, well-organized essay. Make sure you have a clear argument supported by a variety of concrete examples.

A. Racial and ethnic inequalities have been persistent problems in the United States. They have been the focus of periodic waves of both progressive reform and conservative reaction. Discuss public attitudes about racial and ethnic differences and the ways that discrimination and inequality were addressed by politicians, reformers, and other popular movements during the Progressive Era (1890-1919), World War II, and the "Sixties" (roughly the mid-'50s to the early '70s). Keep in mind the range of social and ethnic groups that experienced discrimination during this period.

B. Scholars in many fields claim that the media have transformed American culture and politics since the advent of television. Taking a longer historical perspective, analyze the role of the media in shaping American society between the late nineteenth century and the present. Choose three events or issues, including one from WWI or earlier, and evaluate the role of the mass media—newspapers, journalists, photographers, radio, television, movies, and/or the Internet—in shaping public responses to each. Be sure in your conclusion to address the larger question of the media's role in American culture.

This course is based on three weekly lectures given by a professor, a weekly discussion section, plus readings in a textbook and from primary sources. There is a series of written assignments due throughout the course, including two smaller papers and one longer research paper. The great majority of the questions asked on the exams are derived directly from the lecture themes and the class discussion themes, and to a lesser extent on work done on class papers. Taking good notes and reviewing those notes is crucial to doing well.

Before exams, we go over what the format of the exams will be (identifications and an essay or identifications, quote explication, and an essay). I recommend to my students that they go over their notes and make a list of the identifications, quotes, and essay topics that seem prominent in the lectures and discussions, and then make up a sample exam for themselves and write out the answers. By doing this, the students will effectively review the most important material and also practice the activity of writing in the way called for by the exam—something that helps them avoid choking during the exam.

ANSWERS

B. The Mass Media's Love-Hate Relationship with American Wars

In a time of war, it is very important to have the support of the nation behind the troops and the cause itself. Because the mass media, in the form of newspapers, television, radio, and magazines, reaches such a wide audience, the role of the media in presenting information about the war, and either rallying public support for it or questioning it, is crucial. Throughout American history, journalists have taken positions both in support of and against U.S. involvement in combat. In general, though, the mass media have tended to support American wars at their outset, while turning against them as the cost in lives and dollars mounts. To support this thesis, I will consider the media coverage of the Spanish American War, the Vietnam War, and the Gulf War.

An effective opening paragraph. It provides an appropriate amount of introductory background, then presents a specific thesis—that the media generally begin by supporting involvement in a war, only to object as the cost in lives and dollars mounts. The paragraph ends with a sentence that not only underscores the essay's thesis, but defines the scope of the essay to follow: examples of media coverage in three wars. If you make your assumptions and thesis clear, then tell your reader exactly what you propose to do, you make the examiner's job much easier. Also, by setting out a set of objectives, you make it clear what you are trying to achieve. If the essay does

attain these objectives, it is likely to receive a positive evaluation. If, however, you fail to lay out objectives, you leave the evaluation wholly to the examiner, who may or may not share your assumptions and whose idea of the scope of the question may be different from yours.

The Cuban War for Independence was covered in depth by the Pulitzer and Hearst newspapers during the late 19th century. In their "yellow journalism," these papers sensationalized the brutalities inflicted by the Spanish on the Cubans. These stories of brutality drove public sentiment in favor of the U.S. getting involved with the situation to help the Cubans. Furthermore, after an explosion on the USS *Maine* caused it to sink in Havana Harbor, the papers printed headlines such as "Remember the Maine!" to further push for U.S. involvement.

After the U.S. declaration of war on Spain, and during first three months of battle that ensued, media coverage of the war remained positive. However, when the war shifted to the Philippines, which the U.S. attempted to colonize during a four-year battle, media coverage turned against the U.S. military effort. Many news articles then questioned our motives in the war, as well as the necessity of spending millions of dollars on the fight.

As a point of fact, the student should more explicitly distinguish between the Spanish-American War proper, which was quickly over, and the more protracted action in the Philippines.

In addition, the journalists started to report on the brutality of the American soldiers, relating how they randomly burned the houses of civilians and used water torture to get information from natives. Over time, both the media coverage and the public sentiment toward the Spanish-American War and the subsequent battle in the Philippines shifted dramatically.

A similar shift in media and pubic support of a war occurred during the Vietnam War. Along with newspapers, television was crucial in shaping public opinion toward the war. Early in the war, the mass media presented exaggerated body counts of the enemy, in order to maintain public support for the war. By the later phases of the war, however, journalists reported on the brutalities, such as the Xom Lang massacre, during which American soldiers raped and murdered Vietnamese civilians. Photographs that appeared in both newspapers and television helped spur the growing public opposition to the war. One of the most famous photos showed a naked Vietnamese girl running down the street, burned by napalm.

As a point of fact, the student should make clear that the media did not deliberately distort "body count" figures and other statistics, but, initially at least, reported these uncritically, simply repeating the official numbers without comment or question.

Addressing a different form of the mass media, the comic book, this shift is quite apparent. Using the popular "Ironman" series as an example, the shift is registered in the comics from the mid-'60s to the mid-'70s. In the earlier period, Ironman is represented as anxious to kill the Vietnamese and show off his big American weapons. By the later period, Ironman is much more concerned with bringing peace to the region, and he becomes cynical of the war effort, reflecting the growing public cynicism at the war's end.

In the current era, the media are just as active and important in shaping our sentiment toward U.S. involvement in foreign wars. Our most recent war, the Gulf War, was heavily covered by the media, especially television. Because it was so short, and a seemingly convincing victory for the U.S., media coverage was predominantly positive. Much like the Ironman comics from the mid-60s, the U.S. during this conflict was portrayed as an aggressor with the superior high-tech weapons.

However, only our airborne bombings were covered on television—there was no coverage of the ground war and no reports of brutalities against civilians. Although civilians certainly died in the bombings, these casualties weren't ever focused on because of the brevity of the war. Had the conflict drawn on, however, media coverage of it would certainly have evolved and would perhaps have covered the more brutal and complex aspects of the war.

The first two wars effectively support the essay's thesis. The Gulf War, however, provides more questionable support, since it depends on hypothetical speculation ("Had the conflict drawn on . . .") rather than historical fact. We cannot know how the media would have treated a protracted Gulf War.

By considering the media coverage of the Spanish-American War, the Vietnam War, and the Gulf War, it is clear that this coverage was generally supportive of the military efforts in the beginning of each conflict, but shifted to negative and more complex coverage in the first two Wars after they carried on over time. Of course, it is difficult to say whether the media actually shaped public opinion or simply reflected the growing public frustration with conflicts that dragged on.

I would argue that the media actually play an active role in shaping public opinion, if only because, without their coverage, most Americans would have no idea what exactly was going on in these foreign theaters of war. Because our news media are mass media, and have been for over a century now, the entire country is exposed to the same news coverage. In this way the mass media serve as a unifying force, and the shifting media coverage of U.S. involvement in wars has tended to unite the majority of Americans in their support or opposition to wars at different times.

Overall, this essay is successful because, as with the midterm essay, it has a clear argument or thesis statement, a series of unfolding points that expound on and refine the thesis, a variety of examples and sources to support the points, and a conclusion that recaps the argument and makes a few expansive points. This is the proper length for a 60-minute essay.

The approach to the question is effective. The student takes one aspect of the media, namely the coverage of war, and focuses on that clearly. Weak essays tend not to focus on a specific area and so ramble rather than present a tightly focused argument and evidence for it. Lastly, this is a successful essay for a final in that it builds on material from the whole semester and incorporates this into a longer view on the topic at hand. This is the sort of breadth expected in the final exam.

CHAPTER 18

EMORY UNIVERSITY

HISTORY 131-002: HISTORY OF THE UNITED STATES TO 1877

Margaret Storey, Teaching Assistant

MIDTERM EXAM

MY PRIMARY OBJECTIVES IN THIS COURSE ARE TO GIVE STUDENTS A SOLID UNDERstanding of the central political and social events of the period, to help them develop critical reading and writing skills, and to teach them how respond to questions about history in a cogent, literate fashion.

I want my students to leave the course having gained an appreciation not only of the facts of history, which are vital, but also of the meaning and relevance historical events have for their lives and society. Moreover, I want them to have developed confidence in themselves about speaking publicly and defending their ideas, and to have grown into better writers and communicators.

The examinations test the students on multiple levels. Multiple choice questions quiz them on basic objective facts of history. Short answer and identification questions are generally geared toward themes or features I have highlighted in my lectures. While I do not include essay questions in my exams, my students have ongoing paper assignments throughout the semester. These papers are geared toward the assigned readings and our discussions of them.

I always encourage my students to attend every class and tell them up front that they will only do poorly if they refuse to do the work. It's a hard class, but a straightforward one. If students will attend regularly, do all their work, and come see me whenever they have questions, they will do well.

The midterm and final are important for two reasons. First, they give students a clear structure for study, which is especially necessary for first- and second-year students. Second, they represent one part of a spectrum of tools I use to evaluate students. Not all students perform equally at all tasks. Tests are better for some; class discussions are very good for others; written papers provide still others with the best forum to demonstrate their knowledge and skills.

The midterm accounts for 20 percent of the final course grade.

I. IDENTIFICATION *(each worth 10 points)*

Choose six of the following nine terms or names. In your blue book, clearly note the choice you have made, identify it, and indicate its significance.

John Calvin
Roanoke
headright system
indentured servitude
Townshend Acts
Tea Act
Intolerable Acts
Embargo Act of 1807
cotton gin

ANSWERS

John Calvin

John Calvin was a leading religious figure of the European Protestant Reformation in the early 16th century, in which dissenters challenged the primacy of the Roman Catholic Church. The main tenets of his religious teachings included his belief in justification by faith alone and in predestination. The first tenet was drawn from Martin Luther's rejection of the Catholic Church's endorsement of good works as the way to salvation and emphasized instead that faith in God was all that was necessary for salvation. The second tenet said that certain individuals—the elect—were predestined at birth to be saved and enter heaven, regardless of works or deeds. Both emphasized the omnipotence of God and denied the importance of an earthly intercessor, in the form of a priest, for salvation. The Reformation spawned many religious movements, including the Separatist and Puritan sects whose members came to New England in the 17th century.

For an answer to any identification question to be complete, the student must address the four W's: Who/What? Where? When? Why? The student must demonstrate both knowledge of who or what the term refers to, where that person or thing lived or took place, when in history that person lived or acted, or event occurred, and, most importantly, why it is important or relevant to American History.

The first ID is an excellent example. While it is important to know who John Calvin was, and what he did and when, the answer would not be complete if the student failed to draw the connection between John Calvin's role in the Protestant Reformation and the birth of sects like Puritanism, which so profoundly shaped American history.

Roanoke

Roanoke Island, off the coast of Virginia, was the site of the first English colonization effort in North American in 1585. Sponsored by Sir Walter Raleigh, under a patent granted him by Queen Elizabeth, the first group of settlers included 108 men who were to set up a mining camp in the New World. Within a year of arrival, however, the settlers got into difficulties with the native Roanoke Indians and narrowly escaped attack by escaping onto English pirate ships led by Sir Francis Drake, who had stopped at the colony on his way from the Caribbean. A second attempt at settlement was launched in 1587, this time with men, women, and children sent to establish a farming community on the Chesapeake Bay, north of the original site. The captain of the ship, however, insisted upon landing the settlers at Roanoke, where the Indians living there immediately put up resistance. The captain sailed away again, planning to get reinforcements and return. By this time, however, the English were embroiled in a war with Spain, which prevented Raleigh from sending reinforcements. By the time ships again arrived at the island in 1590, the colonists had disappeared, with only the word CROATOAN —the name of a nearby island—carved into a post. The "Lost Colony" of Roanoke is a perfect example of the real difficulties that lay behind the utopian dreams of men like Raleigh and Drake, who sought riches and fame in the lands of the New World, and spent little time considering the challenges of settling an already-inhabited land.

headright system

The headright system was established by the Virginia Company in 1618 in an attempt to stabilize their struggling colonization enterprise in Jamestown, where the settlers were dying from disease and starvation in large numbers and law and order had virtually disappeared. The measure granted to every settler already in the new colony 100 acres of land and provided that each new settler would receive 50 acres of land. Moreover, every person who paid the passage of a new settler would also receive 50 acres of land as well. The system was designed to encourage the importation of new blood into the colony, and it worked; thousands of immigrants—most of whom were indentured servants—flocked to the new colony from England during the 1620s, giving the settlement a steady foundation upon which to grow and prosper. Without the system of land grants, and the labor system of indentured servitude it encouraged—the colony would most likely have failed.

This is a complete identification. It defines the term in question, provides a reason for the system, and evaluates the results of the application of the system, drawing a definite conclusion.

indentured servitude

Indentured servitude was a system of labor established during England's early colonization period of the 16th and 17th centuries. Typically, a prospective servant would enter into an agreement—the indenture—with another individual, either at home in England or abroad in Virginia, by which that individual paid the servant's passage to the New World in exchange for the servant's labor, generally for a period of seven years. In this way, the planters of the new colony were able to procure cheap labor to farm tobacco, the cash crop of the day. The vast majority of new settlers to Virginia—three-quarters, in fact—were indentured servants. However, their lowly station in life—being for the time of their indenture little more than slaves—made them the most vulnerable to overwork, malnutrition, abuse, and disease. Forty percent did not live to gain their freedom from indenture. The system illustrates the political and economic risks inherent in this early wave of immigration to the New World.

The answer not only demonstrates understanding of indentured servitude, it concludes by demonstrating an appreciation of the historical significance of the institution as an example of the high stakes of New World immigration.

Townshend Acts

The Townshend Acts were a group of Parliamentary Acts that take their name from their advocate, English Prime Minister Charles Townshend, William Pitt's successor in 1766. A man with little comprehension of the colonies or their political or economic needs, Townshend advocated, and got passed, acts providing for new taxes on the lead, paint, paper, glass and tea imported by Americans from England; a suspension of the New York assembly, which had refused to enforce the Quartering Act; and a new Board of Customs to enforce the new taxes with customs agents placed at the colonial ports. As a whole, the acts worked to more completely centralize colonial governance in England, and away from the colonies. They therefore only created greater tensions between Parliament and the colonial assemblies. Protests of the new taxes were begun by citizens who encouraged boycotts of the taxed items and dependence on homemade goods, resistance that eventually led to widespread and united protest throughout the colonies. The most important of these protests included the street riots led by Samuel Adams and the Sons of Liberty in Boston in 1768 and the Boston Massacre of 1769. While refusing to relinquish the Parliament's right to tax the colonies, English legislators finally repealed all but one of the Townshend Acts in 1770—the tax on tea—in an effort to appease the colonists.

Tea Act

The Tea Act, passed by Parliament in 1773, gave the bankrupt East India Company a monopoly on the tea trade to the colonies, and was thereby designed to save the company from financial ruin. On their face, the consequences for the colonists were positive, for the monopoly made it possible for the East India Company to charge less for tea than local American merchants did. The political leadership in the colonies not only understood that the Tea Act would harm American sellers, but also thought that the monopoly would provide a vehicle for further taxation measures. The act, although not in itself very offensive, became the focus of the colonists' anti-taxation protests, which had languished since the repeal of the Townshend Acts in 1770. With Samuel Adams and the Sons of Liberty taking the lead, Bostonians came to the wharf where thousands of pounds of tea was housed, broke open the casks, and dumped the tea into the harbor. The Boston Tea Party, as the protest was dubbed, provoked severe recrimination from England and fueled the revolutionary fervor of the leading anti-taxation resisters.

Intolerable Acts

Officially known as the Coercive Acts, these four laws were enacted at different points in 1774 by Parliament in response to American colonists' ongoing resistance to Parliamentary taxation, and specifically in response to the Boston Tea Party of 1773. The first law—the Boston Port Bill—closed Boston's port until the colonists made reparations to the East India Company for the tea they had destroyed. The second law—the Massachusetts Government Act—turned over all colonial government to royal officials. The third law—the Impartial Administration of Justice Act—dictated that any royal official charged with a crime in the colonies could, if he chose, be tried in England or any other colony than that in which the crime was committed. The fourth law—the Quarter Act—extended the effects of the first Quartering Act, already active in Massachusetts, to all the colonies; it allowed the British government to forcibly house British soldiers in colonists' private homes. The acts were called the Intolerable Acts by the colonists because they appeared to represent an unacceptable level of tyranny on the part of the English government and a desire by that government to enslave the colonists and deny them equal political rights as English citizens. The new laws simply added more fuel to the fire of American revolutionary fervor.

Embargo Act of 1807

Passed by Congress at the encouragement of President Thomas Jefferson, the Embargo Act was a response to the war between France and Great Britain and American desires to trade with either side with no fear of attack from either belligerent. Jefferson proposed that America adopt a plan of peaceful coercion, in which American trading in either British or French ports would be prohibited until the two countries agreed to respect American neutrality and trading rights. The plan, based on the assumption that American goods were deemed essential by both countries, was a total failure. France carried on with the war with apparently little concern for the missing American trade; Britain quickly took over the trading routes America had abandoned. By the end of his term as president, Jefferson had stopped enforcing the law altogether. The Act was repealed early in 1809.

Take care to ensure that short essay responses, while brief, are complete. This one does not stop with defining the Embargo Act, but concludes by evaluating its effect and effectiveness.

cotton gin

For the South, the invention of the cotton gin was the most important product of the Industrial Revolution. Created by Eli Whitney in 1793, this machine removed seeds from the lint of cotton bolls. Traditionally done by hand by slave laborers, the job was so intensive that cotton agriculture had been limited to long-staple cotton, which would only grow in the low-country regions of the South but whose seeds were more easily separated. The type of cotton that would grow in the upcountry South was short-staple and had the prohibitive disadvantage of having many small sticky seeds that were very difficult to remove by hand. The cotton gin took care of that problem and allowed a single slave to de-seed fifty pounds of short-staple cotton a day, as compared to the one-pound-per-day rate by hand. Cotton cultivation spread rapidly into the Southern interior, and the South quickly became the leading cotton-producing region in the world.

Everyone approaches tests differently and should be encouraged to give their full attention to all parts of the test, rather than rushing through one section in order to devote a predetermined amount of time to another part. The most important thing a student can do in this regard, it seems to me, is to immediately read through the entirety of the test before he begins, so that he understands the test's structure and can estimate the amount of time necessary for each part.

II. MULTIPLE CHOICE *(each worth 4 points)*

Read the following questions carefully and circle the appropriate answer for each.

1. Unlike Separatists, Puritans

a. advocated strict separation of church and state
b. practiced passive resistance to oppression
c. remained members of the Church of England
d. were Calvinists

Answer: c

2. The "Wild Bachelors" of Bacon's Rebellion were

a. young merchants upset over high taxes
b. the planter class of Virginia
c. those protesting the importation of African slaves
d. young men frustrated by their inability to acquire land

Answer: d

3. One feature of the American economy that strained the relationship

between the colonies and Britain was the

a. British demand to halt the importation of slaves
b. lack of any British regulations regarding trade with foreign nations
c. growing desire of Americans to trade with nations other than Britain
d. British rejection of the Molasses Act

Answer: c

4. With the British and American victory in the French and Indian War,

a. the American colonies grew closer to Britain
b. Americans now feared the Spanish
c. a new spirit of independence arose as the French threat disappeared
d. the Indians were stopped from ever again launching a deadly attack against white settlers

Answer: a

5. Under mercantilist doctrine, the American colonies were expected to do all of the following except

a. supply Britain with raw materials not available there
b. become economically self-sufficient
c. furnish ships, seamen, and trade to bolster the strength of the Royal Navy
d. provide a market for British manufactured goods

Answer: b

6. The First Continental Congress

a. was attended by delegates from each of the thirteen colonies
b. adopted a moderate proposal for establishing a kind of home rule for the colonies under British direction
c. made a ringing declaration of America's independence from Britain
d. called for a complete boycott of British goods

Answer: b

7. The Northwest Ordinance of 1787

a. provided for the survey and sale of public lands in the Old Northwest
b. established a procedure for governing the Old Northwest territory
c. banned slavery from all territories of the United States

d. cleared the way for ratification of the Articles of Confederation

Answer: b

8. The case of *Marbury v. Madison* involved the question of who had the right to

a. commit the United States to entangling alliances

b. impeach federal officers for "high crimes and misdemeanors"

c. declare an act of Congress unconstitutional

d. purchase foreign territory for the United States

Answer: c

9. The "nullification crisis" of 1832-1833 erupted over

a. banking policy

b. internal improvements

c. tariff policy

d. public land sales

Answer: c

10. The Industrial Revolution in America began in 1791, when

a. the Waltham Plan was designed

b. Samuel Slater set up Brown's Mill

c. shipbuilding boomed in New England

d. shoemakers began using the outwork system

Answer: b

FINAL EXAM

The final accounts for 30 percent of the course grade. The most effective preparation strategy for the exam is to take rigorous notes in class, do all the assigned work, plan to begin studying some time in advance, and ask questions when you need help. Also attend any review opportunities that are given.

I. IDENTIFICATION *(each worth 5 points)*

Choose FIVE of the following terms or names. In your blue book, clearly note the choice you have made, identify it, and indicate its significance.

Free Soil Party
manifest destiny

popular sovereignty
Compromise of 1850
Dred Scott decision
Thirteenth Amendment
Wade-Davis Bill
Freedmen's Bureau
black codes

ANSWERS

Free Soil Party

The Free Soil Party, founded in 1848, was a short-lived political coalition of Northern anti-slavery activists and "Conscience Whigs" who opposed the extension of slavery into the western territorial lands and the creation of new slave states. They nominated Martin Van Buren for president in the 1848 presidential election. The party was a broader one than its precursor, the Liberty Party, which had been solely an abolitionist party; Free Soil was narrower, however, than its successor, the Republican Party, which in 1854 attracted disenchanted Democrats into its ranks as well as Free Soilers.

This identification demonstrates depth of understanding by not only defining the Free Soil Party, but comparing it to its precursor and successor. A good approach to short-answer questions like this is to begin your response by ensuring that you answer the basic question—make the essential identification go on to give something extra, which extends the significance of the original question.

manifest destiny

The phrase "manifest destiny" refers to an idea coined in 1844 by John O'Sullivan, a New York newspaper editor, when he stated that it was the United States' "manifest destiny to overspread the continent" given to the nation by God for the citizens' free development. In practice, the doctrine of manifest destiny came to be understood as a right of Americans to expand the American political, economic, and social systems into the western territories. Proponents of this vision emphasized ideals of individualism, widespread land ownership, and free economic endeavor to justify the taking of lands formerly occupied by Indians as well as the forcible relocation of these Indians when necessary to make room for white settlers. Americans' impetus for expansion was largely rationalized by the idea of manifest destiny, which encouraged small pioneer families to risk everything in search of land and bounty.

While most history instructors these days put decreasing emphasis on brute memorization of facts and phrases, the use of O'Sullivan's phrase here is not only very appropriate, but demonstrates that the student has indeed studied both hard and effectively.

popular sovereignty

Popular sovereignty, also known as squatter sovereignty, was a doctrine established in the late 1840s and early 1850s to support the principle that the people who settled in the as-yet-unorganized territorial lands of the American West should be able to determine for themselves, in democratic referendum when asking to be admitted as a state, whether that state would or would not allow slavery.

If a term has an alternative version, specify it. This identification could have been improved by the addition of a few sentences on the implications and impact of popular sovereignty.

Compromise of 1850

The Compromise of 1850 was a proposal authored by Congressman Henry Clay in an effort to solve a critical political issue that had arisen upon California's request to be admitted into the Union as a free state. At the time, the sectional balance between Northern free states and Southern slave states was equal, with fifteen states on each side. The request for free-state status brought up all the old conflict about slavery that had plagued the Congress since the Missouri Compromise of 1820. Some Southern leaders advocated secession if the western lands were allowed to break the balance between slave and free states. The compromise was written to stave off this eventuality and had the following provisions: California was admitted as a free state; the rest of the land obtained from Mexico in the late Mexican War was divided into two territories—New Mexico and Utah—which would be slave or free based on popular sovereignty, or majority decision, at the time they applied for statehood; Congress abolished the internal slave trade in Washington, D.C., and passed a new and more effective fugitive slave law. The compromise measures were exactly that—none had broad-ranging support from representatives of either section; the slavery issue was muted, but not silenced, by the compromise.

This identification requires a mini-essay, explaining the historical context of the Compromise of 1850 and including preceding compromises and eventual outcomes. The student does well here.

Dred Scott decision

The Dred Scott decision refers to the 1857 decision of the Supreme Court in regard to a lawsuit brought by a slave named Dred Scott against the state of Missouri. Scott had been taken by his master into the Illinois and the Wisconsin territories, both of which prohibited slavery. Scott and his lawyers argued that the state of Missouri should grant Scott his freedom because, by taking him into a free state, Scott's master had made him free. Lower courts had concurred with Scott, but the Missouri Supreme Court had overturned that initial decision. The U.S. Supreme Court heard the case in 1857 and came back with the 7 to 2 decision that Scott was not free, but still a slave. The decision asserted that because Scott was a slave, he had no citizenship rights, including no right to sue in the courts. Moreover, the court also challenged the law that underlay the suit, which was that Illinois and Wisconsin were free areas by virtue of the Missouri Compromise of 1820. The decision declared the compromise unconstitutional because Congress had no power to prohibit slavery in the territorial lands. The Fifth Amendment guaranteed citizens' right to property, which the Congress had violated in the compromise. The decision was vitally important because, although it seemed to have invalidated the political platform of the Republican Party—which opposed the extension of slavery into the territories—the popular effect of the decision was to garner even greater support for the Republican Party and free soilism among citizens in the North. Sectional divisions over slavery were simply hardened, not ameliorated, by the decision.

This answer demonstrates great command of details—yet that never gets lost in the details. The student understands and states the grave significance of the Dred Scott decision.

Thirteenth Amendment

The Thirteenth Amendment (1865) to the Constitution was adopted in the wake of the Civil War and the Emancipation Proclamation. It declared that slavery was illegal in the United States.

While certainly correct, this identification could have been improved by further discussion of the Emancipation Proclamation versus the Thirteenth Amendment. Also, a sentence or two on the immediate effects of the amendment would be in order.

Wade-Davis Bill

The Wade-Davis Bill was proposed in 1864 by two Republicans—Senator Benjamin Wade of Ohio and Representative Henry Davis of Maryland. The bill was a response to President Abraham Lincoln's Proclamation of Amnesty and Reconstruction of 1863, which appeared to them and to much of the Republican Party as far too lenient a plan for dealing with former Confederate citizens and leaders once their states were again under federal control. Wade and Davis proposed instead that former rebel states would be ruled temporarily by a military governor, required half the white males to take an oath of allegiance before the state could form a new constitution, and gave real political power only to those Unionists in the Southern states who could swear they had never aided or abetted the Confederacy. Lincoln, while not outright vetoing the bill, nonetheless thwarted it by letting the time for him to sign it expire. At the same time, Lincoln seemed ready to compromise with the Republicans and attempt a new plan of Reconstruction. Before such a compromise could be met, however, Lincoln was assassinated.

Freedmen's Bureau

The Freedmen's Bureau—formally called the Bureau of Refugees, Freedmen, and Abandoned Lands—was established during the Civil War to oversee the transition of freed people from slavery to freedom, to assist wartime refugees with rations and protection, and to confiscate and manage lands abandoned in the South. The Freedmen's Bureau had agents spread across the South who heard complaints, adjudicated differences between freedmen and former slaveholders, helped make contracts between black laborers and planters, and generally strove to protect the rights of former slaves. It was disbanded under charges of corruption in 1872, much to the disappointment of Radical Republicans, freed people, and bureau agents.

Black Codes

The term "black codes" refers to types of laws, passed by every former slave state after the Civil War, that were designed only to regulate the lives and behavior of blacks, thus establishing a separate legal world for blacks. These laws varied from place to place, but most sought to restrict heavily the freedom of movement and employment for black people, prohibited the presence of blacks on juries, disallowed black testimony against whites in court, required that special licenses be purchased to engage in trade, prohibited the sale or rental of land to blacks, and allowed states to force men and women deemed "vagrants" into plantation labor. The Black Codes were overturned only after the Radical Republicans in Congress took over Reconstruction from the president and required new constitutions in each of the former Confederate states.

II. MULTIPLE CHOICE *(each worth 3 points)*

Read the following questions carefully and circle the appropriate answer for each.

1. The Mormon religion was founded by

a. Europeans
b. Brigham Young
c. Charles G. Finney
d. Joseph Smith

Answer: d

2. By 1860, slaves were concentrated in the "black belt" located in the

a. upper South states of Kentucky, Missouri, and Tennessee
b. Deep South states of South Carolina, Georgia, Alabama, Mississippi, and Louisiana
c. old South states of Delaware, Maryland, Virginia, and North Carolina
d. new Southwest states of Delaware, Maryland, Virginia, and North Carolina

Answer: b

3. The slave Nat Turner led his rebellion in Southampton County, Virginia, because

a. John Brown encouraged him to do so
b. he had spiritual visions which he interpreted as signs to lead a race war
c. his master had falsely accused him of raping a white woman
d. a group of Massachusetts abolitionists promised him freedom and a reward if he succeeded

Answer: b

4. In its effort to impact public opinion, the American Antislavery Society employed all of the following methods except

a. postal campaigns
b. home visits to recruit new members
c. infiltration of Southern slave patrols
d. petitioning Congress

Answer: c

5. What percentage of Southern slaves were employed in industrial enterprises?

a. 5%
b. 10%
c. 25%
d. 60%

Answer: a

6. Southern slavery was characterized by

a. massive mortality
b. malnutrition and bad housing
c. natural increase
d. virtually nonexistent slave families

Answer: c

7. One argument against annexing Texas to the United States was that the annexation

a. could involve the country in a series of ruinous wars in America and Europe
b. might give more power to the supporters of slavery
c. was not supported by the people of Texas
d. offered little of value to America

Answer: b

8. The Wilmot Proviso, introduced into Congress during the Mexican War, declared that

a. Mexican territory would not be annexed to the United States
b. slavery would be banned from all territories that Mexico ceded to the United States

c. the United States should annex all of Mexico
d. the United States should have to pay Mexico a financial indemnity for having provoked the war

Answer: b

9. John C. Calhoun advocated an alternative to the Wilmot Proviso, called the "Common Property Doctrine," which stated:

a. that an amendment to the Constitution should be required to outlaw slavery in the territories
b. that Congress had no constitutional authority to regulate slavery in the territories
c. that the Missouri Compromise line should be extended through the Mexican Territory to the Pacific Coast
d. that the number of slave and free states should always be equal

Answer: a

10. During the debate of 1850, ______________ argued that there was a "higher law" than the Constitution that compelled him to demand the exclusion of slavery from the territories.

a. William H. Seward
b. Henry Clay
c. Daniel Webster
d. Stephen A. Douglas

Answer: a

11. The Fugitive Slave Law included all of the following provisions except

a. the requirement that fugitive slaves be returned from Canada
b. denial of a jury trial to runaway slaves
c. denial of fleeing slaves' right to testify on their own behalf
d. the penalty of imprisonment for northerners who helped slaves to escape

Answer: b

12. Stephen A. Douglas's plans for deciding the slavery question in the Kansas-Nebraska Act required repeal of the

a. Compromise of 1850
b. Missouri Compromise
c. Wilmot Proviso

d. Northwest Ordinance

Answer: b

13. **In "Bleeding Kansas" in the mid-1850s, ________________ was/were identified with the proslavery element, and ________________ was/were associated with the antislavery free-soilers.**

a. Beecher's Bibles; border ruffians

b. John Brown; Preston Brooks

c. the Pottawatomie massacre; the sack of Lawrence

d. the Lecompton Constitution; the New England Immigrant Aid Society

Answer: d

14. Nativists in the 1850s were known for their

a. support of Native Americans

b. support of slavery

c. opposition to old-stock Protestants

d. anti-Catholic and anti-foreigner attitudes

Answer: d

15. The Republicans lost the 1856 election in part because of

a. Southern threats that a Republican victory would be a declaration of war

b. lingering support for slavery in the North

c. Northern opposition to free-soilers

d. the Democrats had presented an excellent candidate

Answer: : a

16. In ruling on the Dred Scott case, the United States Supreme Court

a. hoped to incite further debate on the slavery issue

b. expected to lay to rest the issue of slavery in the territories

c. supported the concept of popular sovereignty

d. reunited the Democratic party

Answer: b

17. The Panic of 1857 was caused by all of the following except

a. a drop in the price of cotton

b. inflation due to the influx of gold in southern California
c. overspeculation in land and railroads
d. overproduction of grain crops

Answer: a

18. The government of the Confederate States of America was first organized in

a. Atlanta, Georgia
b. Montgomery, Alabama
c. Richmond, Virginia
d. Knoxville, Tennessee

Answer: b

19. Confederate batteries fired on Fort Sumter when it was learned that

a. Lincoln had ordered the fort reinforced with federal troops
b. Lincoln had ordered supplies sent to the fort
c. the fort's commander was planning to evacuate his troops secretly from the fort
d. Lincoln had called for seventy-five thousand militia troops to form a voluntary Union army

Answer: b

20. As president of the Confederacy, Jefferson Davis did not exercise the arbitrary power wielded by Abraham Lincoln because

a. of the South's emphasis on states' rights
b. there was such strong agreement on policy in the South
c. he did not believe in strong executive action
d. all of the above

Answer: a

21. In the Civil War, the South won the battle of

a. Vicksburg
b. Bull Run
c. Gettysburg
d. Atlanta

Answer: b

22. The first ex-Confederate state to ratify the Fourteenth Amendment and thus be readmitted to the Union under Congressional Reconstruction was

a. Virginia

b. Arkansas

c. Louisiana

d. Tennessee

Answer: d

23. The goals of the Ku Klux Klan included all of the following except to

a. "keep blacks in their place" through white supremacist ideology

b. prevent blacks from voting

c. keep white "carpetbaggers" and "scalawags" from voting

d. support efforts to pass the Force Acts of 1870 and 1871, which would force blacks away from the polls

Answer: d

24. Congress's impeachment of President Johnson and attempt to remove him from office were directly precipitated by his

a. highly partisan "swing around the circle" in 1866

b. readmission of Southern states under his policies in 1866

c. dismissal of Secretary of War Stanton in 1867

d. advice to Southern states not to ratify the Fourteenth Amendment

Answer: c

25. The Credit Mobilier scandal involved

a. public utilities

b. the Bureau of Indian Affairs

c. railroad construction

d. excise taxes on distilled liquor

Answer: c

CHAPTER 19

FLORIDA STATE UNIVERSITY

AMH 1000: AMERICAN CIVILIZATION

Daniel S. Murphree, Graduate Teaching Assistant

STUDENTS IN THIS COURSE ARE EXPECTED TO BE ABLE TO IDENTIFY CERTAIN PEOPLE, places, and events significant to United States history, to analyze and form their own opinions on past trends and movements, and to evaluate the similarities and disparities between North America's various regions and peoples during the development of the United States. They should improve both their research and writing skills and broaden their ability to interpret different materials through a relatively objective perspective. I hope as well that my students will develop a more complex view of the different issues relevant to U.S. history and be able to ascertain their own personal relationships to these issues. Perhaps most important, students should refine their skills in creating a logical argument and substantiating it with valid evidence.

The course examinations are designed to evaluate mastery of the skills just mentioned. Essay questions test the student's ability to present a clear argument in an orderly and concise format. They are designed to let the students propose any argument they believe pertinent, as long as their argument can be supported by evidence obtained through course lectures, discussions, or readings. Questions are also formulated to test depth of knowledge on certain topics and discourage general or cursory answers. In addition, portions of most questions give students the option to contribute their personal perspectives on the topic in order to help them better understand how their past counterparts reacted to certain issues.

MIDTERM EXAM

Multiple Choice

Directions: Circle the letter that best answers each question. (2 points each)

Questions 1, 6, 10, 12, 14, 18, 19, 22, 24, and 25 are designed to test understanding of concepts discussed in class lectures. Questions 5, 8, and 15 are designed to test understanding of concepts discussed only in assigned readings. Questions 2, 3, 4, 7, 9, 11, 13, 16, 17, 20, 21, and 23 are designed to test understanding of concepts discussed in class lectures and assigned readings.

1. During the American Revolution, Native Americans preferred following which course?

A. Moving to Canada
B. Fighting on British side
C. Fighting on colonists's side
D. Neutrality

Answer: D

2. According to Shawnee Prophet (Tenskawatawa), how should Indians purify their lifestyles?

A. Abstain from drinking alcohol and reject white customs and ideas
B. Move west of Mississippi River and form new agricultural settlements
C. Give up farming and survive only through hunting wild game
D. Eradicate all whites through warfare

Answer: A

3. Which U.S. president signed the Indian Removal Act (1830) into law?

A. James Madison
B. Andrew Jackson
C. William Henry Harrison
D. Jefferson Davis

Answer: B

4. The National Trades Union was:

A. The first U.S. organization promoting temperance
B. An organization of artisans that agitated against unfair British taxes
C. The founding organization of the U.S. Whig party

D. The first attempt at a national labor organization in the U.S.

Answer: D

5. According to your book, what significant event occurred at Seneca Falls, New York, in 1848?

A. First abolitionist society convention
B. First temperance association convention
C. First women's rights convention
D. First Carnival Guild's convention

Answer: C

6. Why did many Americans oppose Horace Mann's new education system created in the Jacksonian period?

A. They disliked the new emphasis on classical education
B. They disliked the new taxes needed to support it
C. They disliked the Whig leaders who ran the system
D. They disliked sending their children to school away from home

Answer: B

7. In 1836, Texas:

A. Gained independence from Mexico and immediately joined the United States
B. Lost a war with Mexico and remained a Mexican territory until 1840
C. Failed to gain annexation to the United States and remained independent for ten years
D. Split in two, with the southern half remaining part of Mexico and the remainder joining the Union

Answer: C

8. What caused the "Mormon War" in 1857?

A. Public discontent over the Mormon's practice of polygamy
B. Indian anger over Mormon expansion into the Utah territory
C. The Utah state government's attempts to remove all Mormons to Wyoming
D. Brigham Young's decision to mine gold in the Black Hills.

Answer: A

9. Why were many Texans outraged after the Battle of Goliad (1836)?

A. Texan troops had fought poorly and surrendered only after a brief fight

B. British troops assisted Mexican forces in defeating the Texans

C. Stephen Austin betrayed the Texans and told Santa Anna about Sam Houston's plans

D. Santa Anna ordered 500 Texas prisoners executed

Answer: D

10. Which statement is not accurate?

A. Most European immigrants of the 19th century preserved their culture by creating ethnic communities within large cities, which had their own churches, newspapers, and schools

B. Most European immigrants of the 19th century experienced a sense of "uprootedness" from what was familiar to them in their European villages and communities

C. Portions of the Alien and Sedition Acts were targeted at European migrants

D. "Birds of Passage" was the term used to describe immigrants who planned on returning to their country of origin.

Answer: B

11. Know Nothings was/were?

A. The term used to describe immigrants who came to the U.S. without an education

B. Early immigrant societies

C. The term European Jews used to describe native born Americans

D. Another name for the American Party, which was a nativist political organization

Answer: D

12. All were nativist organizations except?

A. The Anti-Immigranters

B. American Protective Association

C. Ku Klux Klan

D. The Immigration Restriction League

Answer: A

13. In the Old South, slaves:

A. Were prevented from creating their own cultures

B. Lost contact with their African heritage

C. Created a culture based on African heritage and American customs

D. Refused to incorporate white customs into their culture

Answer: C

14. What was significant about the Stono Rebellion (1739) in South Carolina?

A. Many of the slaves hoped to join and gain the assistance of the Spanish in Florida

B. Many of the slaves hoped to capture the city of Richmond

C. Many slaves received the assistance of Tennessee hillbillies

D. Many plantation owners supplied the slaves with weapons and food

Answer: A

15. According to your book, could blacks own slaves in the U.S. prior to the Civil War?

A. Yes

B. No

C. Happy

D. Sad

Answer: A

16. Which region of the country did the New England Emigrant Aid Society encourage northerners to settle?

A. Texas

B. Kansas

C. Missouri

D. Indiana

Answer: B

17. The Missouri Compromise did all of the following EXCEPT:

A. Abolish slavery in Washington, D.C.

B. Admit Missouri to the Union as a state

C. Admit Maine to the Union as a free state

D. Prohibit slavery north of 36°-30"

Answer: A

18. Following his beating at the hands of Preston Brooks, Charles Sumner:

A. Quickly returned to the Senate so he could resume leadership of the anti-slavery movement

B. Reached a personal reconciliation with Brooks in an effort to alleviate growing sectional tensions

C. Filed criminal charges against Brooks, who served a brief jail sentence for aggravated assault

D. Could not return to the Senate for three years, and his vacant chair came to symbolize Southern brutality

Answer: D

19. Who were the "Fire eaters?"

A. Pacifist Southern abolitionists who saw the Civil War as a humane and righteous cause

B. Militant Northern abolitionists who supported Radical Republican policies

C. Militant western agriculturalists who disliked eastern industrialists

D. Militant Southern nationalist leaders who began to agitate for secession shortly after Lincoln was elected president

Answer: D

20. The Battle of Bull Run in 1861 was significant because it:

A. Showed how well the black troops on both sides could fight

B. Ended Northerners' illusions that the war would be ended quickly

C. Caused the Union to rely more on regular troops than on volunteers

D. Was the only battle to inflict major casualties on civilians

Answer: B

21. The Emancipation Proclamation:

A. Applied only to those areas in rebellion against the Union

B. Was strongly supported by both political parties in the North

C. Applied only to the slaves in Missouri, Kentucky, and Maryland

D. Led European states to grant diplomatic recognition to the Confederacy

Answer: A

22. Despite their differences of opinion, what did most northerners agree on regarding Reconstruction of the South following the Civil War?

A. They all believed Southerners should be punished severely
B. They all believed Southerners should not be punished severely
C. They all believed the Southern states should not be allowed back in the Union
D. They all believed that Southerners should reform their way of life and give up any future plans of succession

Answer: D

23. What was the major accomplishment of the Freedmen's Bureau?

A. Helping freed slaves gain employment
B. Redistributing property to freed slaves
C. Prosecuting former Confederate war criminals
D. Transporting supplies from the North to the South

Answer: A

24. Which of the following was not a reason settlers moved West after the Civil War?

A. Passage of Homestead Acts (1862, 1873, 1877, 1878)
B. Acquisition of territory from Canada
C. Establishment of new territories
D. Mining

Answer: B

25. Why is Turner's thesis on the western frontier inaccurate?

A. Turner emphasizes Indian dominance over Anglo-Americans
B. Warfare in the West did not end until 1925
C. Settlement in the West was neither democratic nor individualistic
D. Turner ignores the contributions of gamblers

Answer: C

Essay Question—*Choose 1 of the following questions and answer it as clearly and fully as possible. (50 pts.)*

Section II of the exam calls for students to choose 1 of 3 possible questions to answer in essay form "as clearly and fully as possible." Students are verbally advised to use the format of the question as an outline for their answers. Essay exam questions are designed to evaluate students' mastery of lecture topics and assigned readings as listed under course objectives.

1. Describe the different policies the U.S. government pursued toward Native Americans prior to the Civil War. How did the policies affect the Indians and how did the Indians respond to them? What were the overall results?

The answer must address the following: names and goals of different U.S. policies, effects of each policy on Indians and examples illustrating these effects, overall results of policies on native societies and United States development.

Students should let the phrasing and structure of the question serve as the format for the answer. Before beginning the essay, review different events affecting native societies, as discussed in course materials. Evaluating these events and their time frames independently should enable you more clearly to recognize the provisions and impact of each policy. This process should lead to an orderly presentation of ideas and provide examples to back up arguments regarding policies.

ANSWER

Prior to the Civil War the United States government pursued a variety of policies toward Indians in North America. Though differing in application and success, all policies sought the ultimate goal of ending native occupation of lands desired by Anglo-American settlers. During the American Revolution both the Continental Congress and British government attempted to gain the allegiance of the different Indian groups or at least ensure their neutrality. Though most natives preferred remaining neutral, as the war progressed the majority ended up siding with the British. The result was ongoing rebel distrust of Indian motives, hostility, and violence. Within Indian communities, the war caused internal dissension among those supporting opposite sides. The split of the Iroquois confederation is an example of this dissension. Throughout the eighteenth century, the Iroquois League was the strongest Indian community in North America. Composed of the Mohawks, Oneidas, Tuscaroras, Onondagas, Cayugas, and Senecas, this group had generally functioned as a unified confederacy in economic and military affairs. The American Revolution changed this. The Mohawks, Cayugas, and Senecas supported the British, while the Oneidas and Tuscaroras supported the colonists. Only the Onondagas remained neutral, and, as the war progressed, their stance caused them to be decimated by both Americans and the British. Though the Iroquois nominally survived after the war, divisions caused by the revolution eventually resulted in disintegration of the dominant Indian alliance east of the Mississippi River. Besides the Iroquois, other groups such as the Creek, Cherokee, and Shawnees also were sucked into the war and shared similar experiences. To different degrees, participation in the American Revolution disrupted life in most Indian communities.

Following the war, all Indians, regardless of allegiance during the war, were treated as conquered peoples. As a result, the U.S. government felt it had acquired the Indians' lands.

Hoping to justify their conquests, the American government enacted a series of treaties in which they allegedly obtained Indian recognition of their claims. Key to the arrangement of these treaties were the participants involved. Often Indian negotiators did not speak for the majority of the Indian community. U.S. officials often negotiated only with pro-U.S. officials. Often such Indians had little power or status among their own people. The result was a treaty process that validated the goals of the U.S. but often ignored the concerns of the Indians. Disgruntled by the treaty encroachments and spurred by the English, many Indians in the Old Northwest unified to resist the Americans. In 1786, Miamis, Shawnees, Delawares, Ottawas, Kickapoos, Potawatomis, and Hurons agreed to cede no more land to the U.S. without approval of all groups. Fearing the growing power of the Confederacy, Washington authorized the raising of a federal army to disperse Indians. Initially, two separate invasions were repulsed by the confederacy until an army under General Mad Anthony Wayne defeated the Indians at the Battle of Fallen Timbers (1794). The Northwest Confederacy dissolved following the defeat, and the Indians were forced to sign the Treaty of Greenville, giving up most of the Ohio country to the U.S.

Despite the general success of the Conquest Policy, late in his second term, Washington adopted a new Civilization Policy for dealing with Indians. This policy was adopted later by Jefferson, Madison, Monroe, and Adams. While still advocating land cessions, the government now attempted to establish a peaceful relationship with Indians by regulating contact between Indians and whites and providing goods to native communities. Rooted in enlightenment philosophy, this policy was based on the idea that Indians basically were equal, but needed help to reach the level of white civilization. To promote this policy, in 1795 the federal government established a series of trading houses throughout Indian territory to curb trader abuses and win native loyalty to the U.S. Through credit obtained from sale of lands, Indians could obtain domestic animals, seeds, and vocational training while also adopting white dress and morals. Under this policy the U.S. gained 70 million acres of Indian land west of the Greenville treaty line. Land was also acquired from Creeks in Georgia, Cherokee in Tennessee, and Sauk and Fox west of Mississippi River.

Indians did not fare as well. In many native communities, traditional power structures collapsed. Old chiefs lost authority because gift-giving power was negated by the factory system. Those chiefs who exchanged land for goods were often attacked by traditionalist Indians. Those groups with no land to sell in the Northwest grew increasingly bitter toward U.S. Rather than U.S. traders, these groups continued to receive assistance from British in Canada (Potawatomis, Miamis, Shawnees). Growing tensions led to growth of a new Indian confederacy organized by the Shawnee Prophet and Tecumseh, a confederacy that allied with British during the War of 1812. British defeat marked the end of unified Indian resistance east of Mississippi River.

By the time Andrew Jackson became president, most whites in the U.S. believed the civilization policy was a failure. In response, Jackson endorsed the idea of removing the eastern Indians west of Mississippi River for their own good. Nevertheless, Indians resisted removal from the very beginning. In 1830, the federal government passed the Indian Removal Act, which called for the U.S. to exchange unsettled territory west of the Mississippi River for Indian lands east of the river. The act appropriated funds for negotiating treaties with the southern tribes and relocating them in the West. Though the Chickasaws, Choctaws, Creeks, and Seminoles were also removed, the most dramatic change involved the Cherokees. Adhering to the Civilization policy, many Cherokees had adopted "civilized" culture of white Americans. Many were educated, owned expensive homes and property, owned slaves, and participated in local western-style governments. Nevertheless, most were forced, often at gun point, to move westward. Between 1838 and 1839, over 15,000 marched and sailed to new lands in Arkansas and Oklahoma. Along the way over 4000 died. By the late 1840s, few large Indian communities remained east of the Mississippi River.

Overall, the U.S. government achieved its goals through the different policies. Indians were removed as obstacles to western expansion. These policies also destroyed many parts of Indian culture and much of the natives' autonomy. Removal also caused future problems with Indians in the West that would exist until the twentieth century.

This essay question evaluates the student's mastery of course objectives under two subheadings: Sub-question 1, "Describe the different policies the U.S. government pursued toward Native Americans prior to Civil War," evaluates the student's ability to know and be able to identify cer-

tain people, places, and events significant to United States history. A successful answer must include the names, provisions, and time frames of the different policies pursued. Important people, groups, and occurrences (Iroquois League, Battle of Fallen Timbers, Andrew Jackson) must be included to demonstrate familiarity with material. Sub-questions 2 and 3, "How did the policies affect the Indians and how did the Indians respond to them?" and "What were the overall results?" evaluate the student's ability to analyze and form an opinion about past trends and movements and to evaluate the similarities and disparities between North America's regions and peoples during the development of the United States.

These questions allow students to express their own views on the different policies and provide suitable examples to back up their arguments. (Example: Civilization policy undermined autonomy in native communities because the factory system diminished the power of traditional chiefs, who lost gift-giving resources.) Subquestions 2 and 3 also prompt students to speculate on different attitudes and priorities held by whites and Native Americans. (Example: Cherokees embraced civilization policy, but whites felt their efforts unsatisfactory and favored removal. Students are also encouraged to determine overall impact of policies in their conclusions based on their own analysis of evidence. (Example: Overall U.S. policies accomplished goals of white society but disrupted native cultures.)

This answer is successful because it addresses all main points in the material covered (different policies and examples of their impact), is presented in an organized manner (different policies discussed as independent units in logical progression), and demonstrates the student's problem-solving skills (analysis of their impact on American society).

2. **Discuss the major events contributing to the sectional crisis between the North and South prior to the Civil War. What were the major issues dividing northerners and southerners? Describe how the differences escalated from debate to violence. Be sure to note all significant occurrences from 1820 to 1860.**

Student answers must include descriptions and impact of the major events of the sectional crisis as discussed in course materials, major issues underlying sectional conflict, and the evolution of conflict from debate to violence.

As with the first question, let the phrasing and structure of the question itself serve as the format for your answer. Begin by reviewing different events comprising the section crisis and the times they took place. Evaluating these events independently should enable you to recognize the evolution from debate to violence, and this should help you to create an orderly presentation of ideas and examples.

ANSWERS

Even prior to the American Revolution, cultural and economic differences divided the peoples of the thirteen original colonies. In 1787, the U.S.

Constitution became acceptable to the states only after compromises regarding slavery and sectional preferences were made (Three-Fifths Compromise). Though few major sectional disputes arose during the late eighteenth and early nineteenth centuries, problems continued to mount until the 1820s, when overall disparities between the North and South surfaced over the most controversial difference, slavery.

Between 1787 and the 1850s, sectional differences over slavery primarily surfaced in the form of political debate. Differences regarding slavery became a national concern in 1820 due to a sequence of events known as the Missouri Compromise. In 1819, the territory of Missouri applied for statehood. Missouri was the first territory west of the Mississippi River to apply for statehood. Until this point, no provisions concerning slavery or admission of states west of the river had been established. Most northerners believed that, due to Missouri's location, it should be admitted under the provisions of the Northwest Ordinance. Southerners believed Congress did not have the authority to prevent the spread of slavery. They wanted to maintain the Senate's balance of free and slave states (12-12) since, by 1820, the House of Representatives was 105 to 81 in favor of the free states. After heated debate, a compromise was reached. Missouri gained admission as a slave state, while Maine, which had separated from Massachusetts a year earlier, came in as a free state. In addition, a line was drawn west from the southern boundary of Missouri at latitude 36'30' to the Rocky Mountains, dividing the lands that would be slave and free.

This compromise functioned until about 1850, when more issues related to slavery came to the forefront. First, in 1849, California applied for admission to the union as a free state. Second, the question of the territory gained from Mexico was still unresolved, and many southerners were agitating for the division of the area into several states. Third, many northerners were agitating for the end of the slave trade in Washington, D.C. Fourth, many southerners were complaining over the poor enforcement of the Fugitive Slave Act of 1793. They wanted a stronger act that would end Northern aid to runaways trying to escape to Canada. In 1850, the Old Whig Henry Clay attempted to resolve all the disputes. Though his compromise measures were defeated when presented to Congress as one law, they were later approved when resubmitted by a New Whig leader, Stephen Douglas. The Compromise of 1850 called for several things: (1) California admitted as a free state, (2) the territorial governments of New Mexico and Utah were able to decide their status by popular sovereignty, (3) the slave trade but not slavery was abolished in District of Columbia, (4) the Fugitive Slave Act was strengthened. Alleged fugitives were denied jury trials, and Northern citizens were ordered to hunt down runaway slaves and turn them in. For a time, this compromise did resolve some of the disputes.

Harmony did not last long, however. In 1854, Illinois Senator Stephen Douglas introduced a bill for organization of the Nebraska territory (including Kansas). Douglas's bill called for revocation of the Missouri Compromise. He proposed that Kansas and Nebraska be organized separately according to popular sovereignty. He tried to convince southerners that, through this method, Nebraska would probably be free while Kansas would probably be slave, thereby maintaining the balance of states. After much debate, the bill eventually passed. Few people in the North or South were happy with the law's provisions. The Kansas-Nebraska controversy basically canceled all earlier compromises regarding sectionalism and slavery. It also prompted a contest between slavery advocates and abolitionists over settlement of Kansas. Each group began encouraging immigration to the region in hopes that they might determine the form of government under popular sovereignty.

After several years of debate and disagreement, sectional disputes transformed into violence. In 1854, the New England Emigrant Aid Society enlisted 1,200 New England settlers to migrate to Kansas. Outraged, thousands of pro-slavery Missourians led by Democratic senator David Atchison, crossed the border in order to counterbalance the anti-slave votes. In March 1855, an election was held to select a territorial legislature. Swamped by Missourians, a pro-slavery legislature was elected. Believing the election was unfair, abolitionists elected a Free-Soil government in a different part of the state. In May 1856, a pro-slavery mob attacked the Free-Soil legislature in Lawrence, Kansas. As revenge, three days later, Free Soilers led by John Brown entered a pro-slave settlement and hacked five men to pieces with swords. Guerilla warfare continued until the Civil War began.

Violence also occurred in Congress. Disturbed over events in Kansas, Senator Charles Sumner delivered a speech condemning and insulting many southern politicians. Two days later, Representative Preston Brooks of South Carolina, a nephew of one of the insulted southern senators, attacked Sumner with a cane, beating him unconscious. Sumner was unable to return to the Senate for four years due to his injuries. Preston was censured, but became a hero in the South, allegedly receiving gifts of hundreds of canes from supporters.

Three years after the massacre in Kansas, John Brown launched an even more daring raid against what he perceived to be slavery interests. In October 1859, Brown and 22 relatives and free blacks attacked the federal arsenal at Harpers Ferry, Virginia. He hoped to incite a general slave uprising throughout the upper South. Brown's force was quickly surrounded by federal troops and captured. Almost half of his men were killed and few slaves joined in his uprising. Tried for treason, he was later executed. Though Brown's raid was unsuccessful, it did inspire more northern outrage against slavery. Many northerners praised Brown's efforts and even advocated further expeditions in the South. Southerners, on the other hand, began to fear slave revolts sponsored by northerners and questioned the U.S. government's willingness to suppress them. Much like the other events, Brown's raid further solidified sectional differences and alienated many southerners from the rest of the nation. The raid also symbolized the change in passions from the spirit of compromise 40 years earlier.

This question evaluates mastery of course objectives under 2 subheadings. Subquestions 2 and 4, "What were the major issues dividing northerners and southerners?" and "Be sure to note all significant occurrences from 1820 to 1840," evaluate the student's ability to identify certain people, places, and events significant to United States history. A successful answer must include the names, provisions, and time frames of the different events. Important people, groups, and occurrences (Missouri Compromise, John Brown, popular sovereignty) must be included to demonstrate familiarity with material. Sub-questions 1 and 3, "Discuss the major events contributing to the sectional crisis between the North and South prior to the Civil War" and "Describe how the differences escalated from debate to violence," evaluate the student's ability to analyze and form opinions about past trends and movements and to evaluate the similarities and disparities between North America's regions and peoples during the development of the United States. These questions allow students to express their own views on the different issues and provide suitable examples to back up their arguments. (Example: John Brown's raid on Harpers Ferry symbolized the change in passions from the spirit of compromise forty years earlier.) Sub-questions 2 and 3 also prompt students to speculate on different attitudes and priorities of northerners and southerners. (Example: Northerners and southerners perceived Brown's raid on Harpers Ferry through totally different perspectives due to sectional priorities and influences.) Students are also encouraged to determine overall impact of policies in their conclusions based on their own analysis of evidence. (Example: Sectional crisis functioned mainly as debate until the 1850s, and then evolved into violence.)

This answer works well because it addresses the main political events and gives examples of their impact, and it is presented in organized manner (the events addressed in chronological order), and also demonstrates significant problem-solving skills in the analysis of the impact of the events discussed.

3. Analyze U.S. expansion westward before and after the Civil War. Define Manifest Destiny and describe its supporters and opponents. How did the U.S. acquire Texas, Oregon, California, and the Southwest? What motivated Americans to move westward following the Civil War?

A good answer will cover the definition, ideology, and reality of Manifest Destiny, methods the U.S. government used to acquire Oregon, California, and Texas, and motivations for U.S. immigration west of the Mississippi River during latter half of nineteenth century.

Again, let the structure of the question provide the structure of the response. Approach migration westward before and after Civil War separately. Formulate a definition for Manifest Destiny before trying to list examples of its impact. Consider the changes the Civil War made on U.S. society before considering the motivations of immigrants after 1865. Recognizing overall changes in the U.S. between periods should enable you to see more clearly the different reasons for immigration and acquisition of lands.

ANSWER

Manifest Destiny was a theory originating in the early nineteenth century that asserted the United States was destined by God and history to encompass the entire North American continent. The theory became popular among many Americans due to the penny press, an assortment of newspapers that emerged during the 1840s and 1850s due to improvements in transportation and communication. Generally politically independent, the penny press employed sensationalism to gain readers and appealed to people of all social classes. Manifest Destiny became a sensational topic. Presented as uniformly supported by all Americans, Manifest Destiny's primary goals were to strengthen the influence of the U.S. and "save" the savages of Texas and the West.

In reality, Manifest Destiny was not uniformly accepted by Americans, and those that did support the theory did not always have selfless motives. Many in the South favored expansion as a way to spread agriculture, slavery, and racial dominance of "inferior peoples." Many in the North opposed it for the same reasons. Democrats favored Manifest Destiny, believing it would reduce the power of the New England states, diminish impulses toward secession and separatism, and contribute to individual mobility and freedom. Whigs opposed expansion due to the political problems arising over expansion of slavery and difficulty of integrating new territories into the union. They also feared their political base would be overrun and moral beliefs would be unenforceable. Perhaps the greatest motivation was economic gain. Southerners saw expansion into Texas and

Mexico as a way to acquire more cotton lands in addition to expanding markets for consumption. They hoped eventually to eliminate or at least severely limit European commerce in Mexico and Central America. Northerners also saw the West as a market for the North's growing commerce in manufactured products as well as a "gateway" to further overseas trade with Asia.

The consequences of Manifest Destiny were displayed in the U.S. acquisition of Oregon, California, and Texas. Since the eighteenth century, the Oregon country had been claimed by Spain, Britain, Russia, and the U.S. By the early nineteenth century, only the U.S. and Britain still had claims there. In 1818, the U.S. and Britain agreed to allow citizens of both countries equal access to the region. Between 1842 and 1845, the number of Americans in Oregon grew from 400 to 5,000. These settlers quickly wrote a constitution and elected a legislature. At the same time British interests diminished as the fur trade became less profitable and fewer British citizens immigrated to the area. While campaigning for the presidency in 1844, James K. Polk challenged Britain, saying that the U.S. must have control of the country up to the 54° 40" parallel or go to war. Angered by his threats, the British refused to relinquish their claims, and war seemed imminent. Realizing the country's unpreparedness for war and growing tensions with Mexico, Polk secretly backed down from his claim and suggested a compromise at the 49th parallel (today's boundary). The British agreed, and the U.S. gained sizeable territory on the northwest coast.

At the same time that the Oregon question was being resolved, American interest in California increased. In the 1820s and 1830s, American traders began moving to California, claiming Mexican citizenship and intermarrying with the local Indians and the Hispanic population. By the 1840s, the region attracted more U.S. interest due to its good harbors and favorable position for Asian trade. In 1846, an "exploration" expedition under John Frémont ventured to California. Though not actually starting hostilities, Frémont encouraged a local group of American residents to rise up against their Mexican rulers. Seizing much of northern California, these residents proclaimed the independent Bear Flag Republic and awaited the outcome of the Mexican-U.S. war then going on in Texas.

The most blatant acquisition in the name of Manifest Destiny came in Texas. Under the Spanish, Texas existed mostly as a sparsely populated borderland between the U.S. and Central America. Hoping to strengthen the area after becoming independent, the Mexican government initially encouraged U.S. settlers to move to the region. Coming mostly from the South and bringing slaves,

U.S. settlers outnumbered the Mexican population by 10,000. By 1836, both the Mexicans and U.S. settlers were dissatisfied with their situation. Disliking the growing U.S. population, Mexico outlawed slavery and halted further immigration. Outraged, local settlers led by Sam Houston rebelled against Mexican authorities. Following a short war with Mexican forces, the rebellious Texans created the new Republic of Texas. Close ties to the U.S., however, eventually led to calls for annexation and further troubles with Mexico. Over the next few years, Texas and Mexico both launched raids over the border, though neither had the resources to cause much damage. Unable to receive much financial support from the European powers, Texas increased its financial ties to the U.S. In March 1845, just prior to leaving office, John Tyler pushed through a joint resolution calling for the annexation of Texas. Unlike a treaty, which needed two-thirds approval of the Senate, a joint resolution only needed a majority approval. As a result, Texas annexation was easily approved. Angered by the U.S. decision, Mexico immediately broke diplomatic relations and sent troops to the Texas border. Shortly thereafter, a skirmish broke out between Mexican and U.S. troops, and Congress declared war. U.S. forces quickly gained the advantage over Mexican troops. By 1847, Vera Cruz, Tampico, and Mexico City were occupied by U.S. soldiers. Because the U.S. had achieved almost total victory, many Americans called for annexation of all of Mexico. Prior to the victories, however, President Polk had sent Nicholas Trist to Mexico with detailed instructions of how to end the war. Though discredited by the president, Trist arranged a treaty based on these instructions, which did not call for the total annexation of Mexico. Under the Treaty of Guadalupe Hidalgo (1848) the U.S. absorbed 75,000 Spanish-speaking inhabitants and 150,000 Indians, and U.S. territory increased by 529,000 square miles, including lands that would become California, New Mexico, Texas, Arizona, and parts of Colorado.

Though the Civil War impeded westward expansion for a short time, the movement quickly resumed after the conflict ended. After 1865, U.S. citizens moved west for a variety of reasons. During the Civil War, Congress adopted a new policy of land distribution in the newly acquired territories. Passed in 1862, the Homestead Act permitted settlers to buy a plot of 160 acres for a nominal fee if they occupied and improved it for at least five years. Initially unsuccessful due to the difficulty of yielding sufficient crops from such a small area, the government revised the act in 1873, 1877, and 1878. These revisions enabled settlers to obtain as much as 1,280 acres of land at lower cost and reduced the amount of improvements needed in order to keep the land.

As a result of the Homestead Acts, several new territories were created in the years following the Civil War. Nevada became a state in 1864 and Nebraska in 1867. Colorado, Arizona, Idaho, Montana, and Wyoming had all reached territorial status by 1870. Seeing the establishment of white-dominated governments in the West, many eastern settlers immigrated, believing the region was becoming safer and more "civilized."

One of the biggest economic attractions of the West was the inception of large-scale mining of precious minerals. Gold was first discovered in California in 1848. A year later, over 80,000 prospectors from the east and Europe immigrated to the area. As the lodes petered out, many of these miners looked for alternate sources. Unfortunately for the miners, most of the accessible gold and silver was gone by the 1880s. As a result, most prospectors gravitated to different regions, often leaving behind ghost towns. Others, however, primarily corporations, took over many abandoned sites and extracted minerals such as copper, lead, tin, quartz, and zinc, materials that had more lasting value due to their importance to developing industries.

Other immigrants were attracted by the vast open lands in the West, particularly the Plains. This region specifically attracted those involved in cattle ranching. After 1865, U.S. cattlemen were attracted by the unfenced and unclaimed grasslands owned by the federal government. By 1866, Texas cattle ranchers were driving close to 300,000 head of cattle per trip to Missouri, where they were shipped east. Though most drives consisted of about 2,000 to 5,000 cows, by 1871, over 1.5 million head of cattle had been driven to distribution centers in Kansas and Missouri. By the 1870s and 1880s, long cattle drives were meeting increasing resistance from westward-moving farmers. Kansas settlers began fencing in lands on the open range obtained through the Homestead Act. Sheep ranchers from California and Oregon began grazing their flocks on range land in Wyoming and Montana. Due to increased competition for lands and the overexpansion of the cattle industry, long cattle drives largely disappeared by the 1890s. In their place, however, many cattlemen established permanent ranches with fenced-in grazing lands. These establishments flourished and caused many of the cowmen to set down roots.

The most influential factor determining settler immigration was the status of Indians in the West. By the late nineteenth century, most of the Indians had been pacified, often through warfare, treachery, and deceit. Consequently easterners saw the West as secure and migrated in greater numbers. From the 1860s to 1880s, the U.S. engaged the western Indians in a number of wars. Though many treaties were signed, many were broken with the end result being the attempted forced movement of Indians off lands desired by miners, farmers, or cattlemen.

This question assesses mastery of course objectives under two subheadings. Subquestions 2 , 3, and 4—"Define Manifest Destiny and describe its supporters and opponents," "How did the U.S. acquire Texas, Oregon, California and the southwest?," and "What motivated Americans to move westward following the Civil War?"— call for identification of significant people, places, and events. A successful answer must include the names, provisions, and time frames of the different policies pursued. Important people, groups, and occurrences (penny press, James K. Polk, Treaty of Guadalupe Hidalgo) must be included to demonstrate familiarity with material. Subquestion 1, "Analyze U.S. expansion westward before and after the Civil War," requires analysis and an ability to evaluate the similarities and differences between North American regions and peoples during the development of the United States. These questions allow students to express their own views on the different issues and provide suitable examples to back up their arguments. (Example: The myth and reality of manifest destiny were quite different.) Subquestion 1 also prompts students to speculate on different motivations for westward migration before and after war. (Example: Motivations prior to war included expanding slavery, the expanding cotton market, and access to Asia, while postwar motivations focused on the quest for minerals, cattle raising, and multi-crop farming.) Students are also encouraged to determine the overall impact of policies in their conclusions based on their own analysis of evidence. (Example: Impact of Indian pacification on settler mentality.)

This answer covers such important points as the main motivations for westward movement during different periods. It is presented in organized manner, the events addressed in logical sequences, and it amply demonstrates problem-solving skills in its analysis of the impact of events on U.S. boundaries.

FINAL EXAM

All students are verbally instructed during the first week of classes on how best to prepare for examinations and research papers. Besides standard reminders concerning the taking of class lecture notes and scheduling of text readings, students are prompted to approach all assignments by considering at least two arguments. When analyzing issues such as U.S. policy toward Native Americans, or public opinion toward the Civil War, students should view the topic from all major perspectives before forming their own personal opinion. The goal is to convince students that much of what they learn about history is interpretation, and, much like history professionals, they should develop an interpretation based on all available perspectives and evidence.

Multiple Choice

Directions: Circle the letter that best answers the question. (2 points each)

Each mulitple-choice question is designed to evaluate the student's knowledge of both general and specific topics covered in the course and determine familiarity with certain themes. Questions 1, 5, 6, 7, 8, 11, 13, 15, 16, 19, 22, 23, and 25 are designed to test understanding of concepts discussed in class lectures. Question 9 tests understanding of concepts discussed only in assigned readings, and questions 2, 3, 4, 10, 12, 14, 17, 18, 20, 21, and 24 are intended to test understanding of concepts discussed in class lectures and assigned readings.

1. In the late nineteenth century, U.S. farmers primarily blamed which three factors for their worsening economic situation?

A. feminists, foreigners, and banks

B. railroads, banks, and eastern manufacturers

C. climatic changes, railroads, and foreigners

D. western industrialists, Nativists, and railroads

Answer: B

2. What were the initial goals of the Populists?

A. Eradicate bank and railroad monopolies

B. Establish stores, banks, and processing plants, independent of middle men, for farmers

C. Eliminate competition from foreign farmers and reduce power of eastern industrialists

D. Form a coalition with Socialists and achieve national political influence

Answer: B

3. Why were foreigners attracted to immigrant ghettoes in the early twentieth century?

A. They sought links to their homeland and protection from discrimination

B. They sought high-paying jobs in the factories located there

C. They sought cheap and easily obtainable plots of land located there

D. They were prohibited by the federal government from settling in any non-urban areas

Answer: A

4. Besides re-emphasizing the U.S. right to oppose European intervention in the

Western Hemisphere, what did the Roosevelt Corollary (1904) proclaim?

A. The U.S. would promote a trade and military alliance with China

B. The U.S. would declare war on any European or Asian country that interfered with governments in Latin America

C. The U.S. would declare an economic embargo on any European country that interfered with the governments of Latin American countries

D. The U.S. would intervene in the domestic affairs of its Latin American neighbors if they could not maintain order themselves

Answer: D

5. The term "White Man's Burden" is best characterized by which statement?

A. The economic debts amassed by white southern governments for supporting their black populations during Reconstruction

B. The United States government's term for describing the Indians west of the Mississippi River

C. The term used by Social Darwinists to describe the lower classes in early twentieth-century U.S. cities

D. The justification used by many white U.S. citizens to christianize and civilize the primitive, colored peoples of the world

Answer: D

6. What was the De Lome letter?

A. A letter written Cuban revolutionaries accusing the Spanish government of human rights abuses

B. A letter written by a Spanish diplomat in Washington, D.C., describing President McKinley as weak and vain

C. A letter written by the Mexican government condemning the U.S. attempt to take Cuba from Spain

D. A letter written by the Spanish monarch condemning U.S. support of Cuban revolutionaries

Answer: B

7. Which of the following was created as a response to the settlement house movement of the early twentieth century?

A. The Department of the Interior

B. Immigration and Naturalization Services

C. The civil service profession
D. The social work profession

Answer: D

8. Moderate proponents of female suffrage in the early twentieth century used which argument?

A. The female's role in the home was not as important as her role in the workplace
B. Since females made up over 50 percent of the U.S. population, they deserved the right to vote
C. Obtaining the right to vote would enable women to better represent their spheres in general society
D. With the right to vote, females would be elected to high positions in the state and national governments

Answer: C

9. What was the Square Deal?

A. Theodore Roosevelt's attempt to reconcile differences between labor and big corporations
B. W. H. Taft's plan for improving U.S. relations with Latin American countries
C. Franklin Roosevelt's initial plan for ending the Great Depression
D. Woodrow Wilson's plan for ending World War I

Answer: A

10. What was the Zimmermann Telegram?

A. A message sent to London announcing the U.S. entry into World War I
B. A German proposal to ally with Mexico against the U.S. during World War I
C. An ultimatum issued by President Wilson in response to German submarine attacks against U.S. shipping
D. A letter in which the Spanish ambassador to the U.S. described President McKinley as weak and incompetent

Answer: B

11. Which of the following was not emphasized by Wilson's Fourteen Points?

A. Freedom of the seas

B. Reductions in armaments
C. Conciliatory treatment of defeated powers
D. Pan-European economic union

Answer: D

12. Why did Republicans oppose Wilson's peace plan for ending World War I?

A. They resented being excluded from the peace process and did not want the Democrats to gain politically in the next elections
B. They believed Wilson's plan was too harsh on Germany
C. They believed Britain and France deserved greater rewards for winning the war
D. They hoped to prolong the war in order to gain total economic dominance throughout the world

Answer: A

13. What was the F.D.I.C. (Federal Deposit Insurance Corporation)?

A. A New Deal organization charged with stabilizing inflation
B. A New Deal Organization created to guarantee all bank deposits up to a certain level
C. A conservative measure enacted to reduce unemployment
D. A liberal measure enacted to increase government involvement in social affairs

Answer: B

14. Which of the following was not a major cause of the Great Depression?

A. Lack of industrial diversification
B. Contraction of international trade
C. Hitler's rise to power in Germany
D. The failure of both large and small U.S. banks.

Answer: C

15 Who advocated redistributing the excess wealth of the rich to the poor masses?

A. Charles Coughlin
B. Huey Long
C. Franklin Roosevelt
D. Sideshow Bob

Answer: B

16. What was the Kellogg-Briand Pact of 1928?

A. A U.S.-sponsored plan to reduce the size and armaments of the world's navies

B. An agreement in which the U.S. agreed to lend Great Britain naval vessels

C. An agreement signed by forty-eight nations that prohibited the use of war as an instrument of foreign policy

D. A three-way defensive alliance signed by Japan, Turkey and Russia

Answer: C

17. The U.S. used nuclear weapons on which two Japanese cities?

A. Nagasaki and Osaka

B. Tokyo and Hiroshima

C. Nagasaki and Hiroshima

D. Hiroshima and Osaka

Answer: C

18. Which option best describes the Truman Doctrine?

A. A policy adopted by U.S. officials to contain Communist expansion by supplying anti-Communist governments with money and military support

B. A U.S. plan for militarily strengthening anti-Communist nations on the Pacific Rim

C. A U.S. initiative to infiltrate pro-Communist regimes in South America

D. A U.S. policy implemented by the Roosevelt administration to counteract growing socialist influence in Canada

Answer: A

19. Created by the National Security Act of 1947, the National Security Council based its operations in which location?

A. The Pentagon

B. The State Department

C. The White House

D. The F.B.I.

Answer: C

20. What event caused the Communist government in Russia to distrust the

United States prior to World War II?

A. U.S. bombing of Kamchatka (1917)

B. Landing of U.S. troops in Siberia (1918)

C. Hoover's repression of Communist party in U.S. (1915)

D. Hoover's desire to acquire Irkutsk oil fields for U.S. (1921)

Answer: B

21. Who was Joseph McCarthy?

A. Senator from Wisconsin who led anti-Communist investigations during the 1950s

B. Senator from New Jersey who sponsored anti-lynching legislation following World War II

C. Senator from Alabama who advocated the use of nuclear weapons in mainland China

D. Senator from California who advocated the legalization of marijuana

Answer: A

22. Which description best describes "de facto" segregation?

A. Forced separation by legal measures

B. Voluntary integration without government intervention

C. Government-sponsored integration

D. Voluntary segregation without government interference

Answer: D

23. In general, Black Power organizations adopted which stance toward race relations in the U.S.?

A. Peaceful demonstration and compliance

B. Reciprocation

C. Forced integration

D. Violent insurrection

Answer: B

24. Why was the Montgomery Bus Boycott significant?

A. It caused Huey Newton to become the symbol of black resistance against white oppression

B. It led to creation of federal amendment outlawing intrastate segregation

C. It showed white Americans that blacks were not content with segregation
D. It provided a good story line for the movie Speed

Answer: C

25. What was the Southern Manifesto?

A. A proclamation endorsed by southern university presidents expressing disapproval of forced integration
B. A speech written by George Wallace in which he promised to never allow black equality in the state of Alabama
C. A statement issued by Martin Luther King expressing his opposition to segregation in Atlanta
D. An agreement signed by 100 southern congressmen in which they pledged to oppose forced integration

Answer: D

Section II of the exam calls for students to choose one of three possible questions to answer in essay form "as clearly and fully as possible." Students are advised to use the format of the question as an outline for their answers. Essay exam questions are designed to evaluate mastery of lecture topics and assigned readings as listed under course objectives.

Essay (50 pts.)—*Answer one of the following questions as fully and clearly as possible.*

1. Compare the role of the United States in World War I and World War II. Discuss public opinion toward U.S. participation in the conflict and the major events leading to U.S. involvement. What were the major contributions of the U.S. in both wars (troops, supplies, money)? What role did the U.S. play in the peace agreements ending the wars?

Good answers must address the following: different roles played by U.S. in both wars, evolution of U.S. public opinion toward wars, major contributions of U.S. and events affected by U.S. participation, and U.S. role in peace agreements.

As with the essay questions in the midterm, begin by using the phrasing and structure of the question as the format for the response. Before beginning the essay, think about the different ways in which the U.S. participated in both wars as discussed in course materials. By evaluating these events and their time frames, you should be able to recognize the differences and limitations of each role. This process should lead to an orderly presentation of ideas and suggest exam-

ples to back up your arguments regarding the policies governing U.S. participation in the wars.

Using the question to structure the response works well with most multi-part essay questions and ensures your addressing all parts of the question.

ANSWER

The United States plunged truly into the international arena with the defeat of Spain in the Spanish-American War at the turn of the century. However, by 1914, America remained mindful of the words of George Washington about remaining clear of foreign entanglements. As the continent of Europe erupted in warfare, the public favored continued isolation and neutrality.

The U.S. involvement in World War I did not come about until the later years of the war (1917). Being separated from Europe by the Atlantic Ocean allowed for many Americans to distance themselves from what was regarded as a European conflict. This mere fact of distance did not, however, suppress America's close ties with the Allied forces, mainly Britain and France. Acknowledging the Germans and Austro-Hungarians as the aggressors, the United States felt compelled to assist the Allies in their defeat. Initially, the U.S. sought to maintain trading rights with belligerents. The U.S. merchant fleet quickly became targets of the German navy and their U-boats. The sinking of these vessels, as well as passenger liners ultimately brought the U.S. into the war.

As an outright belligerent, the U.S. supplied the Allies with munitions and war materiel while its armies were amassed and trained. The influx of new soldiers into the Great War prompted the Germans to sue for peace. An armistice was reached on 11 November, 1918. Following the cease-fire, U.S. President Wilson created a plan for permanent peace known as the Fourteen Points. Wilson's Fourteen Points advocated freedom of the seas, neutral self-determination, an end to secret treaties and, most importantly, a League of Nations. The Treaty of Versailles officially ending the war reflected little of what Wilson had hoped for and was never ratified by the U.S. Senate. Government spending during World War I equaled what the U.S. government had spent on the military during its entire previous existence. With the fighting over, the U.S. once again reverted to isolationism, never joining the League of Nations.

Although, during the years prior to World War II, many Americans harbored the same pacifist sentiments regarding involvement in a European war, the Japanese attack on Pearl Harbor quickly tempered all ambiguity with respect to U.S. participation. Unlike World War I, the United States was attacked directly on its own soil. The attack destroyed nearly the entire U.S. Pacific fleet, except for the aircraft carriers. With its Pacific fleet, the Japanese were able to expand their war gains throughout southern Asia and the Pacific. Although it was Japan who attacked the United States, the U.S. focused its efforts on defeating Germany first.

The landing of U.S. troops in Morocco checked the German advance along the African continent and ultimately led to their ouster from the region. This ensured Germany would not gain access to the Suez Canal and try to occupy India. Victory in Africa also ensured the Germans would not have unrestricted access to the Mediterranean Sea.

Allied control of North Africa allowed for the invasion of Sicily, which led to the assault on and defeat of Italy. The North African and Italian campaigns took two years, during which time the Allies were able to amass an invasion force against German-controlled France. As the western Allies pushed into Germany from the west, Russia pushed in from the east. With both armies closing in and the number of options declining, Germany surrendered in May 1945.

The U.S. involvement in World War II was far greater than it had been in World War I. Like Germany, the U.S. was compelled to fight a two-front war in World War II (only one front in World War I). While the Americans fought alongside British and French allies in Europe, in the Pacific they fought the Japanese virtually alone. The U.S. also became more closely allied with Britain and France prior to World War II due to programs such as Lend Lease. Unlike World War I, World War II ended with large U.S. influence on the final peace arrangements. The U.S. played a more direct role in helping their opponents rebuild after World War II. Through the Marshall Plan, the U.S. not only rebuilt Europe and Asia, but also secured markets for their booming economy. A new peace confederation, the United Nations, was created, in which the U.S. played an active role.

The U.S. hoped to avoid the mistakes it had made after World War I by not becoming isolationist and accepting its role as a world power. While this has proved successful in avoiding another economic depression followed by global conflict, it undoubtedly led to a bipolar world in the form of a Cold War.

Essay question 1 evaluates mastery of course objectives under two subheadings. Subquestion 3, "What were the major contributions of the U.S. in both wars (troops, supplies, money)?" evaluates the student's ability to identify certain people, places, and events significant to United States

history. A successful answer must include the names, provisions, and time frames of the different policies pursued. Important people, groups, and occurrences (Lend Lease, North Africa, Fourteen Points) must be included to demonstrate familiarity with material. Subquestions 1, 2, and 3, "Compare the role of the United States in World War I and World War II," "Discuss public opinion toward U.S. participation in the conflict and the major events leading to U.S. involvement," and "What role did the U.S. play in the peace agreements ending the wars?" call for an analysis of past trends, events, and movements. These questions allow students to express their own views on the different roles the U.S. played in the wars and to provide suitable examples to back up their arguments. Subquestions 1, 2, and 3 also prompt students to speculate on why many Americans did not want to participate in wars. (Example: Many Americans favored the non-intervention stance advocated by George Washington 120 years earlier.) Students are also encouraged to assess the overall impact of policies in their conclusions based on their own analysis of evidence. (Example: U.S. involvement in World War II was essential to victory and helped prevent another worldwide depression, but it also created a bipolar world order.)

This answer succeeds because it covers all the main points required (description of U.S. contributions and degree of involvement), is presented in organized manner (U.S. role covered chronologically, with thorough investigation of each war), and demonstrates problem-solving skills (analysis of how U.S. role differed by war).

2. Describe the major events and figures in the civil rights movement of the 1950s and 1960s. How did the movement change from the 1950s to the 1960s? What methods did civil rights activists use to promote change? What were the successes and failures of the movement? What are the legacies of the movement today?

A successful response includes descriptions and impact of major events and figures in civil rights movement, how the movement changed between 1950s and 1960s, major methods employed by civil rights activists, and failures, successes, and legacies of movement..

ANSWER

The Thirteenth, Fourteenth, and Fifteenth Amendments legally abolished slavery and certain forms of discrimination based on race, but civil rights issues would not be fully dealt with until the 1950s and 1960s. In 1954, the *Brown v. Topeka Board of Education* Supreme Court decision initiated widespread reform. The dispute centered around which school the Brown family would send their child. The family wanted their child (a black girl) to attend a white school close to their home, while the county insisted she attend an all-black school on the other side of town. Thurgood Marshall represented the Brown family in court and argued that the segregated school system separated whites and blacks but did not provide an equal education for both groups. Marshall won his argument by pointing out the many deficiencies in the all-black schools and showing that these schools could not offer blacks the same advancement opportunities as the white schools. The Brown family won the decision, forcing the government to integrate public schools.

In 1955 Rosa Parks joined the civil rights movement by challenging the segregated bus system in Montgomery, Alabama. Rather than adhere to tradition, give her seat up to white passengers, and move to the back of the bus, Parks, a black woman, refused to move when ordered and was arrested. The local black community rallied around Parks and boycotted the Montgomery bus system. Over the next year, blacks refused to ride Montgomery buses and relied on complex car pool systems to move about. Though subject to harassment by local police and white citizens, the boycott succeeded in convincing the bus company to revoke its discriminatory seating policies. More importantly, the boycott showed white Americans that, despite prevailing stereotypes, blacks were not content with segregation and wanted change. It also brought to the forefront of the movement a young black preacher named Martin Luther King, Jr.

The integration of education initiated by *Brown v. Topeka Board of Education* was tested in 1957 at Central High School in Little Rock, Arkansas. When black students attempted to enter the school, the local black mayor had National Guard soldiers block their path. Though not a major proponent of civil rights, President Eisenhower asserted that the federal government would enforce national laws. He nationalized the local soldiers and sent troops from the 101st Airborne division to the high school in order to allow the students' attendance. The local mayor soon backed down, and the high school was integrated. This situation showed that the federal government was willing to use force to promote racial integration in the U.S.

Despite these efforts, not all whites supported integration. In 1956 over 100 senators and congressmen from southern states issued a statement known as the Southern Manifesto. In this statement, these officials promised to take all measures possible to resist government forced integration. Their efforts did help impede integration. By 1965, eleven years after the Brown decision, only 29 percent of southern schools were integrated.

One of the most important outcomes of the civil rights movement of the 1950s was the development of Martin Luther King's non-violence doctrine.

Based on previous measures incorporated by activists such as Mohandas Gandhi and Henry David Thoreau, King advocated passive resistance to violence. Civil rights demonstrators would not strike back against their opponents and would accept arrest and punishment without violent response. This non-violent doctrine was effective, because it showed many whites that blacks wanted equal rights through peaceful means regardless of the costs.

In the 1960s, King and other civil rights activists put the non-violence doctrine to the test. In February of 1960, four black freshmen attending North Carolina A&T decided to protest against a Woolworth's five-and-ten store that discriminated against blacks. The students entered the store in Greensboro, North Carolina, and sat at the lunch counter, where they requested service. Refused service and heckled by whites, the men stayed in the store until closing. The next day they returned with fourteen more students. Similar sit-ins spread throughout the country. Over 70,000 people participated in sit-ins at restaurants, libraries, and train stations. Though results were mixed, sit-ins did gain exposure for the civil rights movement and showed the effectiveness of the non-violence doctrine.

In addition to sit-ins, many protesters engaged in freedom rides. In order to make the federal government enforce court decisions prohibiting discrimination on interstate buses and trains, black and white volunteers boarded several buses whose routes passed through Alabama and Mississippi. They expected opposition and hoped this would instigate the government to take action. In 1961, buses passed through Alabama, where many riders were beaten and forcibly removed from the vehicles. When a federal marshal trying to protect the riders was also beaten, Attorney General Robert Kennedy deputized local officers as federal marshals and ordered them to escort the buses to Mississippi. Though the riders were arrested in Mississippi, federal pressure caused most states to accept desegregation of buses and trains.

Encouraged by these efforts and hoping to get Congress to enact civil rights legislation, King and other activists organized a peaceful march on Washington D.C. in August 1963. Far exceeding their expectations, over 200,000 people assembled on the mall in front of the capital. Capped by King's "I have a dream" speech, the march was televised nationally, further heightening support for equal rights.

Though the movement was growing, many civil rights proponents believed more action was needed. Some believed the key to success was utilizing the black vote. In order to do this, restrictions on black voting, many stemming from Reconstruction, had to be overcome. In the summer of 1964, hundreds of black and white volunteers, many of whom were college students, traveled to Alabama and Mississippi to register black voters and open schools devoted to improving literacy and teaching black history. Most faced stern opposition from whites. Hundreds of volunteers were beaten and arrested, and six were killed. Nevertheless, 60,000 blacks were registered, and the event showed America that discrimination could be defeated, even in the South.

Combined, these actions throughout the 1950s and 1960s finally yielded change. In April 1963, JFK presented to Congress legislation prohibiting segregation in stores, restaurants, theaters, and hotels and barring discrimination in employment. Due to Congressional opposition, the bill was not passed before JFK was killed. The next year, however, these measures, known as the Civil Rights Act of 1964, became law. In addition, a year later, LBJ succeeded in passing the Voting Rights Act through Congress. This law suspended literacy and other voter tests and authorized federal supervision of registration in voting places. With these acts, all government-authorized racial discrimination in the U.S. ceased to exist.

The civil rights movement continued in limited form after the 1960s and still gains support today. Two main legacies still promote debate in the U.S. today. In addition to other civil rights legislation, LBJ enacted the new policy of affirmative action in 1965. Under this policy, government employers, as well as many private organizations, adopted the practice of actively recruiting and hiring minorities as compensation for past injustices. Today this policy still provokes disagreement. Many also still debate whether segregation still exists, despite the gains of the civil rights movement. Though the Civil Rights Acts ended "de jure," or legal segregation, "de facto" segregation (segregation by practice) still exists. These legacies promise to keep the cause of civil rights in the forefront of public debate for the near future.

This question assesses mastery of course objectives under two subheadings. Subquestions 1, 3, and 5, "Describe the major events and figures in the civil rights movement of the 1950s and 1960s," "What methods did civil rights activists use to promote change?" and "What are the legacies of the movement today?" call for identification of people, places, and events significant to United States history. A successful answer must include the names, provisions, and time frames of the different events. Important people, groups, and occurrences (Rosa Parks, Non-Violence Doctrine, Freedom Rides) must be included to demonstrate familiarity with material. Subquestions 2 and 4, "How did the movement change from the 1950s to the 1960s?" and "What were the successes and failures of the movement?" require analysis and formation of opinions

backed by suitable examples. Subquestions 2 and 4 also prompt students to speculate on different attitudes and priorities held by whites and blacks (Example: Thurgood Marshall's argument for integration versus white politicians' resistance based on tradition as seen in the Southern Manifesto.)

This answer covers the major required points and is presented in an organized manner, with the events addressed in proper chronological order from early 1950s to the late 1960s. There is an adequate analysis of the civil rights movement and impact on present society.

3. Describe the reform efforts of the Progressive Era. Define Progressivism and explain why people supported the different attempts at solving the problems of U.S. society. Describe the main areas of reform (political, social, other?) and the changes that took place. What role did Socialism play before World War I? Analyze the successes and failures of the movement and the long-term effects still evident today?

An adequate answer will define Progressivism, discuss its ideology, and identify its supporters. It will further describe the various areas of Progressive reform, the role of socialism, and the failures, successes, and legacies of movement.

ANSWER

Progressive reform of the early twentieth century stemmed from public unrest over poverty, government inefficiency, and overall inequalities prevalent throughout the U.S. during the latter half of the nineteenth century. Progressivism was a democratic reform movement supported by a wide variety of individuals who worked for an improved democratic process, the regulation of business, social justice, and progress. Progressives believed overall that society was capable of improvement and that the U.S. was destined to grow and advance on a number of fronts. Progressives approached reform with a businesslike attitude, hoping to improve society through efficiency and organization. Unlike many groups that blindly agitated for change, Progressives established certain goals which were pursued through organized measures. Progressives did not adhere to Social Darwinism. They believed that people failed not due to their inherent lack of fitness for survival but due to the effects of an unhealthy environment. They believed government should play an important role in rectifying this problem. Those supporting Progressive reforms did not all belong to one single social group. Few people supported all Progressive goals, though most people supported at least one. Most people saw the need for reform but did not always agree with the Progressives' stand on reforming particular issues.

Under this wide umbrella of people and ideas, several reform movements stood out. Hoping to improve crowded living conditions in industrial cities, many educated members of the middle class helped establish a series of settlement houses located throughout the country. The "settlement house," a term borrowed from similar structures in England, was a community center established to help immigrant families adapt to the language and customs of the U.S. Staffed usually by young women fresh out of college, over 400 settlement houses functioned during the early years of the 20th century. Among other things, these centers provided care for children of working parents, health clinics, and job training for immigrant men. Most importantly, settlement houses precipitated the creation of the social work profession.

At the local, state, and national level, Progressives instigated a variety of political reforms. During most of the nineteenth century, city government was dominated by political machines with little input from most citizens. By 1900, citizens again wanted to participate in the running of their communities and attempted to gain control from the political bosses. Their most successful effort was gaining the adoption in many cities of city commission and city managers. City commissions, non-partisan, elected officials, replaced city council members that were often appointed by political machines. City managers, professionally trained administrators with no interest in politics, were preferred by Progressives to mayors because mayors tended to respond only to the dictates of the political bosses. By the 1920s, hundreds of cities throughout the country had adopted the new government format. Many Progressives believed that government at the state level was no less corrupt or inefficient. Rather than letting unqualified officials determine issues important to the citizens, many reformers attempted to gain more direct control of legislation. In several states, they succeeded in enacting four major reforms.

1. Initiatives allowed reform groups to submit legislation directly to the voters without interference from the legislature.
2. Referendums enabled the electorate to approve or disapprove of actions by the legislature through a direct vote.

3. The direct primary enabled voters to select their candidates rather than have them chosen by party bosses.

4. The recall gave voters the right to remove office holders if enough signatures were obtained by petition calling for a special election.

Even at the national level, Progressives promoted change. Progressive ideals inspired national office holders, primarily Theodore Roosevelt and Woodrow Wilson. Progressive condemnation of corruption among the major political parties caused the parties to lose their widespread support in the U.S. Though Democrats and Republicans remained the major parties, the percentage of voters supporting them steadily declined. Voter turnout declined from 81 percent in the late 1800s to 59 percent in 1912. Progressive reforms began to work outside of the party system and formed the first organized lobby groups. Not adhering to party lines, these groups supported any candidate who favored their ideas.

Many Progressives, like their counterparts in the early nineteenth century, saw alcohol as the root of many social problems. Most temperance advocates were working-class women, concerned with the use of their husbands' paychecks for drinking, and factory owners concerned with drunkenness on the job. Formed in 1873, the Women's Christian Temperance Union included 245,000 members by 1911. This group enlisted church organizations and anti-immigrant supporters to promote their cause. Following the end of World War I, temperance advocates had gained enough support to propose an amendment to the U.S. constitution. In 1920, the Eighteenth Amendment, outlawing the production or sale of alcoholic beverages went into effect.

Because a large majority of women supported reform movements, many soon joined together to assert their own rights. Early proponents for the female right to vote based their arguments on natural rights. They asserted that women deserved the same rights as men and disputed the idea that women's only role in society was as mother, wife, sister, or daughter. Radical for its time, this argument promoted widespread opposition from both men and women who believed females should primarily be concerned with family. More moderate proponents of suffrage advocated a different argument and gained more support. Rather than enabling women to escape their traditional role, moderates preached that obtaining the vote would enable women to better represent their "spheres" in general society. Gaining the vote would allow women to support causes beneficial to the family— temperance, settlement houses, poverty relief.

Organizing as the National American Women Suffrage Association, this moderate plank had over two million supporters by the end of World War I. In 1920, they succeeded in having passed the Nineteenth Amendment, guaranteeing full political rights to women in the U.S.

Supporting many of the same reforms as the Progressives, socialists in the U.S. gained their highest popularity during the first decade of the twentieth century. Led by Eugene V. Debs, the Socialist Party gained political support concurrently with the Progressives. In the 1912 presidential election, Debs received over one million votes. Socialists also held over a thousand local and state positions throughout the country. Socialism appealed both to intellectuals and the lower classes. Debs received votes from both urban immigrant laborers and southern and midwestern farmers. Prior to the Russian Revolution, many Americans saw Socialism as a viable alternative, no more radical than Populists of Progressives.

Perhaps more than any other reform group in U.S. history, Progressives as a whole achieved a greater degree of widespread change in a variety of areas. Though Progressivism gradually died out, Progressive ideals inspired American leaders until World War I and still guides reform efforts today.

The question consists of two major parts. Subquestions 1, 2, 3, and 4, "Describe the reform efforts of the Progressive era," "Define Progressivism and explain why people supported the different attempts at solving the problems of U.S. society," "Describe the main areas of reform (political, social, other?) and the changes that took place," and "What role did Socialism play before World War I?" call for identification of certain people, places, and events significant to United States history. Important figures, groups, and occurrences (settlement houses, Eugene Debs, Prohibition movement) must be included to demonstrate familiarity with material. Subquestion 5, "Analyze the successes and failures of the movement and the long-term effects still evident today" requires analysis of historical material and allows the student to express his or her own views on key issues, furnishing suitable examples to back up these views. (Example: Progressive reforms had more widespread impact in more areas than any previous reform movement.)

The answer included here is a good one because it provides a necessary definition or Progressivism, identifies the movement's key figures, and discusses its accomplishments. The response is presented in an organized manner, with different kinds of reform addressed in different sections, and it demonstrates the student's problem-solving skills through its analysis of the overall effectiveness of Progressive reform.

CHAPTER 20

THE UNIVERSITY OF IOWA

HISTORY 16B:61, AMERICAN HISTORY: THE UNITED STATES SINCE 1877

Colin Gordon, Associate Professor

TWO EXAMS, A MIDTERM AND A FINAL, ARE GIVEN IN THIS COURSE. THE MIDTERM requires paragraph-length essay-type answers, and the final calls for a combination of paragraph-length answers and a longer essay. The exams account for 40 percent of the course grade. The midterm accounts for 15 percent, the final 25 percent. Some semesters, only a final is given, accounting for 40 percent of the course grade.

MIDTERM EXAM

View the course as an introduction to major themes in modern American history. Since I teach the 1877-present component of this course, I strive to make connections between historical and contemporary problems and issues.

1) You have one hour. You should devote roughly ten minutes to each of the five short answers.

2) Write the questions in any order that suits you, but ensure that your answers are clearly labeled.

3) Use only the exam books provided, cross out rough work, do not rip out pages.

4) Relax.

From the list below, pick any five (5) topics. For each, write a thoughtful, well-

illustrated paragraph which identifies, establishes historical context, suggests historical importance, and (where appropriate) provides illustrations or examples.

Liberal internationalism
"Jim Crow" laws
the family wage
The Great Uprising of 1877
social Darwinism
The Compromise of 1877
The Dred Scott case
nativism
the great migration
dollar diplomacy

ANSWERS

1. Liberal Internationalism

"Liberal internationalism" was the political and economic philosophy best captured by Woodrow Wilson in the World War I era. Liberal internationalism tied the pursuit of free trade, free markets, and free political institutions into an overarching view of world politics and political development. In the Wilson era, this was partly an effort to distinguish the United States from the conventional colonialism of its European allies, and partly an effort to distinguish American economic policy from the sharply contrasting world view offered by the emergence of the Soviet Union in 1917. Liberal internationalism remained only an idea, as Wilson's plan for a League of Nations fell apart in 1920. Liberal internationalism re-emerged in the 1930s, offering barriers to trade as an explanation for the Great Depression. In some respects, Wilsonian internationalism was realized in the Bretton Woods system after 1945.

Remember, in this exam, each of the five questions answered is of equal importance. Plan your time accordingly.

2. The family wage

"The family wage" is the economic and political and social assumption that men are "breadwinners" and women are naturally domestic creatures. This idea was commonly used to depress wages for "women's work." It was used by many in the labor movement (the AFL and the CIO) to argue for higher wages for men. And it was used by many during the Depression to argue for the illegitimacy of women's work—which was (in this view) depriving male breadwinners of means of supporting their families. The family wage was often written into public policy, as in the Social Security Act of 1935, which created two distinct tracks of social assistance: the first a "family wage" model of contributory social insurance, the second a much more meager program of aid for single mothers.

Focus your response by including the key terms in your first sentence. While keeping your response tightly focused, show mastery of the course material by bringing in directly related concepts and issues.

3. Social Darwinism

Industrialization exaggerated class divisions in society and generated a range of political demands for reform or social assistance. A prominent argument against social assistance in the late nineteenth century was the idea, first popularized by Herbert Spencer, of "social Darwinism." Social Darwinism applied the biological theory of "survival of the fittest" to the social and economic order. In this view, those who succeeded did so because they adapted successfully to the demands of the modern economy. As in the parallel "gospel of wealth," those who were rich deserved to be rich. By contrast, those who suffered under industrialization were being "weeded out." The implication, which would survive in conservative economic philosophy through the twentieth century, was that any effort to redress economic inequality went against the natural working of the economy.

In defining a historical idea, explain the idea, mention related or competing ideas, and summarize the actual impact of the idea.

4. "Jim Crow" laws

In the wake of the Civil War, and more directly in the face of the threat of agrarian populism in the 1880s and 1890s, many Southern states began enacting laws which segregated public spaces and restricted black participation in politics. While Southern states were nominally bound by the Civil War amendments to the Constitution, the Supreme Court decision in *Plessy v. Ferguson* (1896) allowed them to build "separate but equal" facilities for black and white citizens.

Jim Crow laws included the segregation of public spaces such as bus stations, public accommodations, and movie theaters. And they included restrictions on suffrage such as poll taxes, literacy tests, and the "grandfather clause" (which exempted those whose grandfathers had been free, that is whites, from meeting the new requirements). The persistent Jim Crow laws were the primary focus of the modern civil rights movement after World War II.

According to Prof. Gordon, each short answer should accomplish the following:

- **It should define or describe the topic at hand.**
- **It should establish the historical context.**
- **It should argue the historical importance.**
- **It should, where relevant, provide appropriate examples or illustrations.**

5. Nativism

Nativism was a response to the demographic upheaval of the late nineteenth century, especially the waves of immigration from Eastern and Southern Europe and the migration of African Americans into the urban North. Established immigrants, mostly of British or German heritage, keenly resented and feared the influx of those who were of different religions, nationalities, races, and languages. Nativism formed the foundation for many anti-immigrant groups and campaigns, including the Red Scare of 1920, the rise of the second Ku Klux Klan in the 1920s, and the passage of sweeping restrictions on immigration in 1924. Just as importantly, nativism lurked behind a range of other reforms and anxieties in the Progressive era and after. Much of the Progressive era concern for the health and sobriety of urban workers was tinged with the belief that those workers were not yet "American."

FINAL EXAM

The final calls for a combination of paragraph-length answers and a longer essay. The exams account for 40 percent of the course grade. The midterm accounts for 15 percent, the final 25 percent, but, some semesters, only a final is given, accounting for a full 40 percent of the course grade.

Professor Gordon comments that he looks for his students to develop at least three major abilities as a result of taking this course:

- **The ability to put contemporary issues in historical perspective.**
- **The ability to think critically about the past.**
- **The ability to read primary documents and to understand them in their historical contexts.**

1) You have two hours. You should devote roughly one hour to each part of the exam.
2) Take the time to outline and organize both your short answers and your essays: a short well-crafted answer is always better than a long and incoherent one.
3) Write the questions in any order that suits you, but ensure that your answers are clearly labeled.
4) Use only the exam books provided, cross out rough work, do not rip out pages.
5) Relax.

The exam allows the student to choose five short answers and one essay question. All of these must be answered.

Part I : Short Answers

From the list below, pick any five (5) topics. For each, write a thoughtful, well-illustrated paragraph which identifies, establishes historical context, suggests historical importance, and (where appropriate) provides illustrations or examples.

The Taft-Hartley Act (1947)
Reaganomics
The Truman Doctrine
the New Left
Watergate
Montgomery Bus Boycott
Congress of Industrial Organizations (CIO)
open shop movement
the politics of growth
Social Security Act

The final exam is not an exhaustive test of knowledge. Since the exam allows substantial choice of short and essay questions, students can and should focus their pretest studying on a few key problems.

ANSWERS

Don't get lost in the details. Keep your attention on the big picture: What political choices are made and why? What options or choices are not made and why? In any analysis or argument, you should stress the importance of the historical themes or episodes you are describing—and not just the fact that they happened.

1. The Taft-Hartley Act

The Taft-Hartley Act was passed in 1947 in an effort to roll back many of the gains made by the Wagner Act of 1935. The Wagner Act had been a controversial focus of business and conservative ire since its passage. During the reconversion debate of 1945 and 1946, many argued that such relics of the Depression and the New Deal could and should be scrapped. This reflected both a sense that such programs were no longer needed (that economic growth would take care of everything) and the atmosphere of postwar anti-Communism. Taft-Hartley forced CIO unions to sign anti-Communist affidavits. A number of major unions, numbering one million members, refused and were expelled. Taft-Hartley also restricted picketing and boycotting, and allowed states to pass "right to work" laws, and many states, especially in the South, did so.

Don't worry about the details. It is more important to have a grasp of the basic issues and of how they change over time.

2. The Truman Doctrine

The Truman Doctrine was major speech in 1947 designed to win Congressional support for foreign aid to Greece and Turkey. Congress had proven generally reluctant to finance American leadership of the Bretton Woods system, and the Truman administration increasingly realized that major programs of foreign aid would be necessary. Truman's tactic in 1947 was to shift Congressional attention to the "Communist menace" by portraying the economic troubles of Greece and Turkey as part of a large and nefarious Communist threat. This effort to "scare the hell out of Congress" worked, and the administration was able to follow up with the Marshall Plan. In the long run, the Truman Doctrine set the harsh tone of the Cold War and laid much of the ideological foundation for McCarthyism.

In short answers, avoid description. Instead, wherever possible, link the short answers to larger issues. For example, the answer on the CIO (which follows) is careful to mention the New Deal context in which new labor organizations was possible.

3. CIO

The AFL was a national umbrella organization of skilled workers, representing only a fraction of the American workforce. When the Depression hit, the limits of the AFL became glaringly apparent. Even as the New Deal looked more and more favorably on collective bargaining through 1933-1935, the AFL dragged its feet. In response, a few major unions—led by John Lewis's United Mine Workers—bolted the AFL and formed the new Congress of Industrial Organizations in 1935. The CIO, responding to the Wagner Act of 1935, was dedicated to a far more exclusive system of labor relations and focused its energies on the core of mass production industry: automobiles, rubber, and steel. The CIO won a series of landmark strikes (most notably the Flint sit-down of 1936-37) and emerged as a major force in national politics and especially the Democratic Party. The CIO's efforts not only organized millions of workers, but also pressed the New Deal to a much more consistent defense of collective bargaining rights.

While I do not expect essays or short answers written under exam conditions to be works of literature, it is nevertheless important to organize and present answers as clearly as possible. I do not expect a blizzard of details; I do expect a carefully reasoned argument, illustrated appropriately with historical examples.

4. Politics of growth

After World War II, the country faced three choices in economic policy: a radical renewal of the New Deal, a conservative abandonment of the New Deal, or a middle path based on sustained economic growth—"the politics of growth." The politics of growth emerged because the argument for the New Deal was undermined by the end of the Depression and the newfound prosperity of the postwar era. At the same time, the argument against the New Deal was undermined by the importance and necessity of a government role in the economy through the war and into the Cold War years. In this sense, the New Dealers conceded on the growth of the welfare state and put their faith in private employment. The conservatives conceded a government presence in the economy. And both embraced postwar internationalism as the key to continued growth and prosperity. The "politics of growth" lasted (and paid off) into the late 1960s.

For the short answers, it is important to establish the historical importance and context of the event or person being discussed.

5. Montgomery Bus Boycott

In 1955, a small act of civil disobedience—the refusal of Rosa Parks to relinquish her bus seat to a white man—sparked the modern civil rights movement. There had been very little progress on civil rights since the Supreme Court struck down "separate but equal" in the Brown decision of 1954. In Montgomery, and elsewhere, civil rights activists began to follow through on Brown, employing community action, organized boycotts, and nonviolent civil disobedience to highlight the core issues of the civil rights struggle. In Montgomery, a local grass roots organization (mostly women) joined forces with an emerging national movement led by Martin Luther King, Jr. to challenge the assumptions and practices of segregation in the South. Montgomery was followed by numerous similar protests, including the "freedom rides" though segregated bus stations and the "freedom summers" of voter registration drives. This phase of the civil rights movement culminated with a March on Washington in 1963 and the passage of the Civil Rights Act in 1964.

Students should divide their time equally between the short answers (5, worth 10 percent each), and the essay (worth 50 percent).

Part II: Essay Question

Choose one (1) of the questions below. Answer the question in such a way that you draw (where relevant) on both lecture material and course readings. Ensure that your essay both argues for the historical importance of the critical period you have identified and sketches the historical pattern before and after.

When precise directions are given, read them and heed them.

1. **What was the most important event, episode, period, or development in the growth of the federal state after 1870?**

2. What was the most important event, episode, period or development in the history of American race relations after 1870?

3. What was the most important event, episode, period, or development in the changing status of women in the United States after 1870?

4. What was the most important event, episode, period, or development in the history of American labor relations after 1870?

5. What was the most important event, episode, period, or development in the history of American foreign policy after 1870?

6. What was the most important event, episode, period, or development in American electoral politics after 1870?

To perform well in the course and on the exams, keep up with assigned reading. More importantly, do the reading consistently through the term rather than cramming at the end. In this way, you will be able to make better sense of the lectures and gradually develop an appreciation of the basic themes of the course.

In answering an essay question, it is important to sketch out your answer in order to ensure that you cover the appropriate time period, and that you are able to illustrate each significant event or era.

ANSWER

4. The passage of the Wagner Act and the formation of the CIO in 1935 marked an important landmark in the history of American labor relations. For the first time, the Wagner Act clearly specified the right to bargain collectively.

Begin with a clear thesis statement—something that can be supported by the evidence you build into the rest of the essay.

American workers suffered tremendously through the rapid industrialization of the late nineteenth century. Their efforts to organize were routinely frustrated by a very limited sense of their collective and individual rights. Employers, politicians, and the law all privileged property rights over bargaining rights and viewed workers only as individuals. When workers' organizations threatened "order" or the "rights of property" they were summarily crushed—most famously in the great railroad strikes of the 1870s and 1880s.

It is important to touch upon the following:

- **the conditions of industrialization and the character of the nineteenth-century working class**
- **obstacles to unionization or organization and the labor violence of the 1870s and 1880s**
- **changing notions of labor and industrial democracy through the Progressive and World War I eras**
- **the decline of labor and the rise of the "open shop" in the 1920s**
- **the impact of the Depression, labor's response to the New Deal (and vice versa)**
- **the Wagner Act, as passed in 1935 and amended in 1947**
- **labor and the "politics of growth" after 1945**
- **labor and economic decline after 1965**

Even the Progressive era did little to advance workers' rights—indeed, many Progressive reforms such as welfare capitalism were meant to encourage loyalty to employers and discourage solidarity with other workers. When Progressives spoke of industrial democracy, they meant little more than the improvement of industrial conditions to the extent that workers would be happier and less likely to strike. In the wake of World War I, the rhetoric of industrial democracy collapsed into a series of bitter postwar strikes (most notably the great steel strike of 1919), which rolled back many wartime gains.

The 1920s were marked by further union losses and the growth of the "open shop" movement. The open shop was articulated as an alternative to a "closed shop" in which workers all belonged to a union. With this rhetoric, employers were largely successful in selling the idea that the "open shop" preserved individual rights. Labor, on the other hand, argued (with little effect) that the closed shop reflected the classical principal of democratic majority rule. A union elected by a majority of workers represented all workers. The law remained with employers, who went so far as to employ "yellow dog" contracts by which union activity was grounds for dismissal.

A good answer must be carefully organized, and it must develop an argument which covers the entire time period under consideration. It must demonstrate mastery over course material by providing appropriate historical examples or illustrations.

The economic crisis which began in 1929 transformed American labor relations. Labor increasingly understood the importance of secure representation. And politicians increasingly understood the importance of an economic policy which would both garner workers' votes and bring about economic recovery.

The first glimmer of reform came in 1933, when the National Recovery Act included a section on collective bargaining rights. Employers interpreted this to mean that proprietary "company unions" would fit the bill, but workers were able to press the administration and the courts to a defense of "real" collective bargaining.

The NRA was thrown out in 1935, but was followed by the Wagner Act, which firmly established the right to collective representation and bargaining. For workers, this was the fruit of a long struggle. For the New Deal, it was the only avenue for reform and recovery left after the debacle of the NRA. For most employers, the Wagner Act was an unwelcome threat. With the AFL's reluctance to follow up on the Wagner Act, a group of industrial unionists formed the Congress of Industrial Organizations (CIO) to organize mass production industry.

The CIO counted a number of impressive early gains in the automobile and steel industries, but ran into tougher sledding in smaller firms and the industrial South. The Second World War gave the CIO the opportunity to solidify its gains, but it also slowed new organization or growth. In 1947, the Taft-Hartley Act rolled back much of the Wagner Act, both by forcing CIO unions to submit anti-Communist affidavits and allowing states to draft "right to work" laws which re-established (especially in the South) the "open shop." Labor could do little more than sign on to the "politics of growth."

While some workers did quite well in the postwar system, those outside the core CIO industries saw few gains. And once postwar growth slowed in the late 1960s, the entire postwar compromise began to fall apart. Today, organized labor represents about the same percentage of workers as it did in the early 1920s.

Although its gains were limited by the war and postwar restraint, the rise of the CIO in response to the Wagner Act of 1935 marked a turning point in American labor relations. For the first time, American workers had an unambiguous right to bargain collectively through representatives of their own choosing. This led to rapid growth of the labor movement and the emergence of the CIO as an important political power.

CHAPTER 21

UNIVERSITY OF MICHIGAN

HISTORY 160: U.S. HISTORY 1607-1865

Jason M. Barrett, Instructor

MIDTERM EXAM

THIS EXAM IS DESIGNED TO ASSESS HOW WELL YOU'VE BEEN KEEPING UP ON, AND thinking about, the course material from the first half of the term. Keep track of the time, write simply and directly, and plan out your answers. Also, please do not write in pencil, and please do write on only one side of a page. We will keep time on the blackboard, and call out when the exam is half over, and when there is five minutes left. Good luck.

PART I: ID's *(30 minutes)*

For four of the six terms listed below, please provide the following information: a) *who* or *what* the term indicates; b) when and where the term is important in American history; and c) why the term is *significant* (what larger issue does it highlight, illustrate, or explain?)

Virginia Company of London
Black Robe
Royal African Company
Bacon's Rebellion
Non-separating Congregationalism
George Whitefield

ANSWERS

Virginia Company of London

Established in London, in 1606, along with the Virginia Company of Plymouth, by King James as a joint-stock company with a monopoly on settling the North American mainland between present day New York and South Carolina (the Virginia Company of Plymouth was given privileges in New England, present day Connecticut, Massachusetts, New Hampshire, Maine, etc.). Established Jamestown on the Chesapeake in 1607/8, first permanent English settlement in N. America.

Demonstrates the profit-seeking nature of early American exploration and settlement. The colonization of Virginia was, first and foremost, big business. This tells us a great deal about the early settlers' motivations, and helps account for their responses to the challenges they found in the New World: especially their resistance in the first years to abandon the search for riches and establish farms.

Read the question carefully, then be certain that you provide the information the question asks for. Note that these short-answer questions clearly address the *who* or *what*, the *when* and *where*, and the issue of *significance*—the three items the question asks for.

Black Robe

"Black Robe" was the name given by Native Americans to Jesuit (Catholic) missionaries in the New World. The name derives from the Jesuits' clothing, which consisted of black robes. The Jesuits were active in French areas of colonization, particularly in present-day Canada along the St. Lawrence Seaway; and their efforts spanned the seventeenth and eighteenth centuries.

The Black Robes contrast with both the Spanish and the English missionary efforts. The Spanish missionaries were, on the whole, more brutal in their attempts to convert natives to Catholicism. In contrast, the Puritans of New England made relatively weak efforts to convert the native population. The Black Robes put less emphasis on the natives' intellectual understanding of Catholic doctrine, and more emphasis on their faith in salvation and submission to Jesus and the Christian God. Most importantly, the Black Robes tried to merge Catholic doctrine with native culture. In all these comparisons, we can see a range of European attitudes toward the Native American populations, and a variety of ways in which the European "civilizing mission" was carried out.

Successful students do not worry about facts and dates; they try to integrate the material they already have into coherent wholes. They come up with explanations of relationships between issues and come into the exam with a few ideas that explain the relationship between economics and religion, or the problems of political ideology, or race and ethnicity. Odds are, you can make at least one of your ideas fit into one of the questions offered.

Royal African Company

Established in 1619 as the Company of Royal Adventurers, and transformed into the Royal African Company (RAC) in the mid-17th century. King James authorized the company with a monopoly charter, which gave the company the sole right to trade slaves to English possessions in the new world (Virginia and the English isles of the Caribbean). Based in London, the RAC worked to build or occupy forts along the West African coast, where RAC agents could deal with African slavers from the interior who brought their "goods" to the coast. In 1698, the RAC lost its monopoly, and the slave trade boomed under free competition.

The fate of the RAC demonstrates the importance of entrepreneurial and free market factors to the success of the slave trade. Once the RAC's monopoly was broken, there was a dramatic rise in the volume of the slave trade—and a corresponding rise in the productivity of the New World plantations. The introduction of competition gave rise to more and better-built ships, a drive into the interior of the African continent in search of slaves, economic efficiency in the trade, and a transformation from the agency method of sale to commission merchandising.

Mr. Barrett stresses three major competencies he looks for:

1. **Mastery of the course material**
2. **Critical and original thinking about the course material**
3. **Clarity of expression**

Bacon's Rebellion

In 1676 Nathaniel Bacon led a rag-tag "army" into the Virginia frontier on a series of raids against the native tribes. But what started out as a war against the natives with marked racial overtones quickly turned into what Edmund Morgan calls a "civil war": Bacon's Rebellion was turned against the governing powers of the colony and succeeded, for a time, in splitting the gentry class into two factions at war with each other, dissolving any claims to legitimacy by either side. Within a few months, English ships arrived to quell the rebellion, and Governor Berkeley regained control of the colony. He carried out retributive trials and condemnations, and by the close of 1676 the king's peace had been restored.

Bacon's Rebellion illustrates both the racial and class-based factors which encouraged the transformation to an African slavery labor system. Racial hostility toward the natives sparked the rebellion, but it quickly turned into a class conflict of poor Virginians against the gentry. With the introduction of African slavery, the gentry were able to deflect a good deal of that class antagonism into racial antagonism against Africans. This gave white Virginians a sense of racial solidarity, and defused the class conflict, which might otherwise have erupted again.

PART II: ESSAY *(30 minutes)*

Choose one of the two questions below, and write a four- to six-paragraph response. Organize your essay clearly, and be sure to make specific references to the course material.

A. Compare the development of New England and Virginian societies from settlement to the Revolution. How or why did these colonies, both settled largely by British immigrants, evolve into such different cultures? In your answer, you may want to consider religious beliefs, economic institutions, political culture, race and ethnicity, and/or the law.

B. Pretend you are a colonist in 1774; write a letter to relatives in England explaining your views on the brewing imperial crisis. You can choose whether to be a loyalist, a patriot, or "on the fence," and you can be a merchant from Philadelphia, a poor farmer from Georgia, a wife and mother from Massachusetts, or any other occupation/location you want. BUT be sure to indicate who/what you are in the course of the letter. Discuss *specific* events and ideas of the crisis that have grabbed your attention, and be sure to explain *why* you feel the way you do about a possible revolution. You can make citations to the course material in the margins, if you don't want to break character.

ANSWERS

A. It is curious that two areas settled by people who came from one country could develop such different cultures. But when you consider how very different New England and Virginia were by the time of the Revolution, I think what is even more interesting is that they came together to fight for a single cause: the United States. The most important differences between them were their economies and their politics. But you have to begin with why each came to the new world in the first place. I will discuss these topics and then argue that two events—the great awakening and the Stamp Act crisis—caused Virginians and New Englanders to start thinking of themselves as Americans.

The main idea of the essay is clear within the first two sentences. The rest of the introduction explains the topics to be discussed and their connection to the essay's main idea.

Even though they came from the same place, the first British settlers were very different, and came with different purposes. As Lockridge talked about, the Brits who settled in Massachusetts were mainly middle-class, and many were dissenters back home. In Dedham, the settlers tried to establish a kind of religious utopia, where the focus was always on God: they founded the town with a church compact. In Virginia, Morgan says, the original settlers were mostly upper-class, and they came to strike it rich by discovering gold or spices. Jamestown was like Dodge City, especially after John Smith was sent home. In Virginia, the main goal was to become rich and go back home.

Cites course material (Lockridge and Morgan); uses specific examples (Dedham, Jamestown, John Smith); and makes a clear point that is directly related to the essay's main idea.

The paragraph would benefit from one more sentence that draws a direct comparison between Virginia and New England.

The main problem people in New England faced was that they couldn't grow any crops that would sell well back home. Shipping became a big industry in New England, but most were subsistence farmers isolated from the outside world. As we talked about in lecture, Massachusetts's economy was split between these two types: merchants and fisherman vs. western farmers. But both, in their own way, were the result of the original mission. Merchants thought they were living examples of God's providence, and rural villages thought they were keeping their church pure. In Virginia, the settlers did not discover gold, but they did discover tobacco—which was probably better. Everybody in Virginia grew tobacco—it was a "cash crop economy" (Morgan). Eventually, Virginians stopped wanting to get rich and go back home. They wanted to get rich and stay in Virginia. William Byrd is a perfect example of this: he built his mansion for permanence (lecture slides).

Again, cites course material and uses specific examples; but also builds upon the paragraph above, and attempts to explain why these economies developed as they did.

The politics of each were very different, too, and were a result of the original mission of each and the economies that developed around them. Massachusetts was a pretty democratic place, at least for the 17th century. Lockridge explained the town meeting system, and how it allowed most of (white/male) town members to participate. Dedham definitely had its "town fathers," who wielded most of the power, but the expectation was that they were to work on consensus; they couldn't do just *anything*. The town fathers were usually the elders in the church, too. The system worked to keep outsiders away, and maintain the town's purity. In Virginia, things were almost the opposite. There, the gentry ran the colony as a private club. Small farmers were dependant on them to get their crops to market (Morgan). The gentry appointed themselves as justices of the peace (lecture), which meant they pretty much ruled their counties like dukes or earls.

Lacks specific examples for New England (would like to see a particular person named as a "town father" or a particular example of consensus rule) and is weighted toward Dedham, to the neglect of Virginia. But in the discussion of Virginia, the student does a nice job of building on the economics outlined in the paragraph above.

With these colonies so completely different from one another, what is truly amazing is that they actually came together to fight a revolution to become a single country. Two events, I think, were most important for causing this. The first is the Great Awakening. Ben Franklin described in his autobiography how news of Whitefield's tours would travel up and down the coast. People in Charleston and Boston all responded to his sermons. The new religion gave the colonists their first common identity, or cause. And Parliament actually helped the colonists band together. Before the Stamp Act, they had mostly passed laws that affected one colony or region. The Dominion of New England is an example of this (lecture): it only affected New England. But the Stamp Act affected all the colonies, so people in Williamsburg thought they were in pretty much the same boat as people in Boston. They began to see attacks on one as attacks on all of them, and that's why they banded together in 1776. So, even though Virginia and New England were very different in their purposes, economies, and politics, they actually wound up having a common interest: independence.

I think this is a great conclusion because it challenges the premise of the original question. Also you can tell the student has been thinking about the Great Awakening and the Stamp Act crisis in broader terms than were laid out in lecture or the texts. A more sophisticated comprehension of the issues would probably disprove his contention, but this line of reasoning clearly shows independent, critical reading and thinking.

This essay is representative of what my very best students commonly achieve. Perhaps one or two students per term can write and think this clearly about the course material. I would give this essay an A. Even though there are minor points that could be tightened, and the overall thesis is not especially insightful or original, it does show a solid comprehension of the course material and an ability to integrate a variety of sources of information into a comprehensible narrative. It is *very* clearly organized and does an excellent job of citing course readings. Without the last paragraph—which is the one place where the student demonstrates original and critical thinking—this might be an A-.

FINAL EXAM

This exam is designed to assess how well you've kept up on, and thought about, the material from the second half the term; and also how well you've integrated it with the material from the first half of the term. Keep track of the time, write simply and directly, and plan out your answers. Also, please do not write in pencil, and please do write on only one side of a page. We will keep time on the blackboard, and call out when the exam is half over, and when there is 10 minutes left. Good luck.

Note that the instructor has indicated the amount of time that should be devoted to each section. Questions are worth one point for each minute indicated; that is, identification and short answer sections are worth 30 points each, while the long essay is worth 60 points.

PART I: ID's *(30 minutes)*

For four of the six terms listed below, please provide the following information: a) *who* or *what* the term indicates; b) *when* and *where* the term is important in American history; and c) why the term is *significant* (what larger issue does it

highlight, illustrate, or explain?).

Federalist #10
Dred Scott
the "American System"
Temperance
Seneca Falls
Charles Finney

ANSWERS

Federalist #10

Essay written by James Madison in 1787/88, which argued for the ratification of the Constitution. In it, Madison explained the way electoral districts would work under the new federal system and is most famous for its proposal of large—rather than small—districts to make the system work. It was printed in New York newspapers.

Madison's ideas about districts radically changed the way people had thought about republican governments. Before them, people thought republics could only be small and pretty homogeneous. Madison thought of a way a big, diverse country could still be republican. This was one of the most important principles of the Constitution.

Dred Scott

A slave who lived in Missouri with his master Sanford, who was an army officer. Sanford was stationed in Illinois for a while and took Dred with him. This was in the 1840s. While in Illinois, some abolitionists convinced Dred to sue for his freedom—claiming that he couldn't be a slave because Illinois didn't allow slavery. The case went all the way to the Supreme Court in 1857.

The Dred Scott case helped push the U.S. into the Civil War. The court ruled that Illinois had to respect Sanford's property, i.e., Dred, so he remained a slave (actually, some abolitionists bought his freedom). Northerners were outraged at the decision, and the South was outraged at the North's outrage. It helped create the crisis that caused the war.

The "American System"

This was the name given to Henry Clay's policies for America. There were four main issues: protective tariffs, internal improvements, high prices for western lands, and the Bank of the United States. These were the main platforms of the Whig party in the second party system (1830s and '40s), and the main disputes between them and the Democrats.

Clay's American System illustrates the Whig vision of America. The Whigs wanted progress and development; they saw an expanding and commercializing republic. The tariffs would help American industry, internal improvements would speed transportation of goods to market, western land prices would discourage speculation, and the Bank would keep currency sound. The Democrats opposed such "progress" because they saw it as corruption.

Examine the *parts* of the question. Make certain that your response clearly and explicitly addresses *each* of those parts.

Temperance

This was the movement to discourage drinking of alcohol in the Jacksonian era (1830s and '40s.) It was spread throughout the U.S., but was usually strongest where there were evangelical religious groups or hostility between immigrants and natives. Temperance advocates thought drinking led to vice, and they wanted to clean up the country by convincing people not to drink.

Temperance is an example of the ways in which Jacksonians believed in progress. They really thought—at least those who were in reform movements like temperance— that if they just got rid of drinking, then poverty, prostitution, and crime would disappear. It also shows how Americans disagreed about just what constituted progress: anti-temperance people thought the reform movements were trying to make them conform to middle-class values.

PART II: SHORT ESSAY *(30 minutes)*

Choose one of the two questions below, and write a four- to six-paragraph response. Organize your essay clearly, and be sure to make *specific* references to the course material.

A. Andrew Jackson and Henry Clay each thought that he was the true heir to the principles of the Revolution, and that the other was a corrupt imposter. Who was right? Explain the Democrats' and Whigs' political platforms and ideologies, and how each attempted to build on the principles of the Revolution.

Would Thomas Jefferson or the other revolutionaries agree with the Whigs' and Democrats' ideas about the republic?

B. What effect did the market revolution have upon the roles of women in the early republic? What was the market revolution? What opportunities or challenges did it present to women? In your answer, you may want to think about the following: the Second Great Awakening, Lowell, Massachusetts, reform societies, middle-class culture, and/or republican motherhood.

Mr. Barrett suggests this strategy for approaching essay questions:

1. **Read the question and determine what it is asking you to write about. (This isn't always obvious.)**
2. **Sketch a brief outline of your response.**
3. **At the beginning and end of each paragraph, glance over your outline and make sure you are still on track—that you're still writing about the topic raised in the question.**
4. **Before you write the last paragraph, quickly reread the introductory paragraph. Does what you are about to write address the same issues? (It better!)**

ANSWER

A. Jackson and Clay

Andrew Jackson and Henry Clay were bitter political rivals. Jackson led the Democrats, and Clay led the Whigs. These were the main parties of 1830s and '40s. Their biggest dispute was over the Bank of the United States. But they also fought over internal improvements and tariffs. Both the Democrats and the Whigs thought they were the ones who upheld the principles of the Revolution, because they each thought that their plan was the best one for the country.

This introduction suggests what the main topics are going to be, but it's not clear from this what the 'main idea' of the essay will be. Also, it's evident that the student hasn't quite gotten the point of the question, which was about Jacksonian *and* revolutionary ideology.

Alexander Hamilton proposed the first Bank of the United States in the early 1790s. It expired before the War of 1812, and a new one was created after the war. That bank's charter was going to expire in the 1830s, and that's what caused the parties to debate it (Watson). The Whigs were for the Bank because they thought it helped keep the currency stable. The Democrats were opposed to it because they thought it created an unfair advantage for the wealthy. So, the

Whigs thought the Democrats were out to destroy the key to prosperity, and the Democrats thought the Whigs were out to keep their rich friends happy.

Most of this is accurate, in a general way, but the examples used do not really address the main issue: Whig and Democrat views on banking. More importantly, this paragraph does not connect to anything else in the essay. There's no clear point.

Another issue was internal improvements. The Whigs thought bridges, canals, and toll roads were a big part of America's future, because they helped more people get involved in the market economy (Watson.) The Whigs wanted to spend government money to help build these internal improvements. The Democrats thought that it was unfair that some of Maine's tax dollars would be spent on a harbor project in Alabama. Again, they thought the Whigs were just out to help their rich friends who would get the contracts. Whigs thought the government should be active, and the Democrats thought it should be passive (lecture).

No specific examples here; however, the answer does give an accurate, though very general, description of the parties' views. But it does not expand upon what's in the preceding paragraphs.

The Democrats and Whigs also disagreed about tariffs. The Whigs wanted to help American industry, like the Lowell textile mill, so they proposed a tariff to raise prices on imported goods. This hurt farmers, who had to pay higher prices, so the Democrats opposed the tariff. The Whigs thought that farmers ought to want to help American industry, because "they were all in it together" (lecture); but the Democrats, again, thought the Whigs just wanted to help their rich industrialist friends.

The example here is not relevant to the main topic: tariffs. It repeats—for the third time—the same Democrat catchphrase for Whigs. It is basically accurate, but does not explain the issue is any depth. Again, it's not clear what this has to do with the purpose of the essay.

I don't think either the Democrats or the Whigs really saw things like the colonists who fought the Revolution. And I think Jefferson would agree with me. Jefferson thought too much of individual rights to agree with the Whigs' "we're all in it together" attitude, and as president he was too activist for the Democrats. I think the Whigs and Democrats lived in a very different world than the America of the Revolution, so it only makes sense that they would see things differently.

This is the first time the student has addressed the question's main issue. This would have been a good way to organize the essay, but it's only offered at the end and has little to do with the body of the essay. It does show, however, that the student *has* thought about the issue.

This essay is representative of what the upper third of my students can achieve. I would give this essay a B. It is very clearly organized and does a nice job of summarizing the course material on Jacksonian politics—though in a very general way. It attempts to cite course material, but is repetitive and sometimes vague. It is clear this student has done at least some of the reading and has come to lecture, and has taken the time to memorize the parties' main platforms. But it is not evident that he has *thought* about the issues raised by these policies, or that he's been able to integrate material from different sections of the course (for example, the Revolution). Most importantly, this essay lacks a central idea or issue to tie it together and give it coherence.

Part III: Long Essay *(60 Minutes)*

Choose one of the two questions below, and write a 6-8 paragraph response. Be sure to make *specific* references to the course material, and choose issues/events that span the entire course.

A. Were Americans *more* religious, or *less* religious, in 1865 than in 1607? Why? What currents or events in American history contributed to the rise—or fall—of religious belief? In your answer, you may want to think about the following: puritanism, the Half-Way Covenant, the First Great Awakening, evangelicalism, Revolutionary disestablishment, the Second Great Awakening, reform societies, and/or science and technology.

B. Some historians have described American history from settlement to the Civil War as one long utopian experiment. Do you think this is true? Why or why not? From the Puritans to the patriots to the temperance advocates, were Americans (and colonists) realists, or were they dreamers trying to achieve a perfect society? In your answer, be sure to make *specific* reference to course material, and discuss "dys"topian elements of Americans' attitudes as well as utopian.

ANSWER

A. Religion

I think Americans were probably just about as religious in 1865 as in 1607. Religion had changed a lot, but I think they believed in it just as much. Religion was always an issue in American history, from the Puritans to the evangelicals. But many events like the Half-Way Covenant and the market revolution caused religion to change. American religion became the thing that caused reform societies like abolitionism to grow up in the Jacksonian period (1830 to 1860). That's what makes it so important.

What is going to be the 'main idea' of this essay? It mentions some topics, but it's not clear if or how they will be addressed in the essay. The organization of the essay is equally obscure.

The Puritans were very religious. They came to America because they couldn't practice their religion the way they wanted to in England. The Puritans were mostly congregationalists, who believed that each congregation was autonomous. They were also Calvinists. They thought the church was the most important part of the town (Lockridge), and the wealthy people usually ran the church. In colonial Massachusetts, the government was supposed to force people to go to church. They had to pay taxes to support their churches. It was like everything was focused on God.

This is generally accurate, but much too vague. It lacks specific examples, and the effort to reference course material is weak. It's clear that the topic of this paragraph is 'Puritanism,' but what about Puritanism? What is the main point here?

But then people gradually stopped believing in the church so much, which is why the Half-Way Covenant happened. They were trying to get people to come back to church. But when the Great Awakening got going, everybody flocked back to hear sermons. Jonathan Edwards was a very popular preacher in the Great Awakening. He traveled all over the country, and huge crowds came out to see him. Ben Franklin went to see him, and even Ben—who wasn't very religious at all—wanted to give him money. America was really religious during the Great Awakening, even though people argued about it all the time. Like the New Lights and the Old Lights.

The organizational principle here is apparently chronology, but it's not clear how we got here: *Why* did people stop going to church? The discussion of Edwards has mistaken him for George Whitefield (the preacher Franklin heard). The description is otherwise generally accurate. Again, what is the point of this paragraph? What is it *doing* in this essay?

One of the things we talked about in lecture was that part of the Revolution was about disestablishing the state churches. Even though the Constitution says "one nation under God," the Americans who wrote it weren't very religious. They didn't like churches. But then the Second Great Awakening happened, and everybody went back to church. This happened in the 1830s. This time, though, the people participating in the awakening were evangelicals. They believed in going out and converting people. This is one of the reasons why they created missionary societies—to help convert natives. The people in the Second Great Awakening were very religious. People in the first awakening just wanted to be left alone to practice their own religion. People in the second one not only wanted to be able to believe what they wanted about religion, but they wanted you to believe it, too! So I think Americans were more religious in 1865 than in 1607, they wanted everyone to believe like they did.

You have to cut the last paragraph on a two-hour exam a lot of slack, but this does not seem to cohere with the rest of the essay. Indeed, it contradicts what was said earlier. The discussion of the Constitution is dramatically overstated. It does not address what has been discussed in the body of the essay.

This essay is representative of what the majority of my students achieve on exams: the average. I would give this essay a C. It does offer rough summaries of a few of the issues from the course, but they are chosen haphazardly and lack specific examples or references to the course material.

The paragraphs clearly contain single topics, but it's not clear what the connection is between them. The introduction lacks a clear organizing idea, and you can watch the essay's coherence disintegrate paragraph by paragraph after that. This student probably knows much more than he's been able to express, but he cannot organize his thoughts clearly and succinctly in a timed exam. Such organization is the biggest hurdle students face in the timed exam situation. Given limited time, it may be difficult to devote a portion of that time to sketching an outline of your answer. Difficult, perhaps—but an essential step to producing a coherent response that is more accurately reflective of what you know.

CHAPTER 22

THE OHIO STATE UNIVERSITY

HISTORY 151: AMERICAN CIVILIZATION SINCE 1877

Richard M. Ugland, Academic Program Coordinator/Lecturer

ONE OR TWO MIDTERMS (ONE IF A TERM PAPER IS ASSIGNED) AND ONE FINAL EXAM are given in this course. Usually, each midterm counts for 25 percent of the final grade. The midterm and final exams are designed to assess the following:

- mastery of a broad range of information concerning the people, forces, and events that have shaped the development of the United States
- ability to think critically about historical issues
- written communication skills, which are essential to the study of history

In addition, to perform well on the examinations, students must demonstrate an ability to draw conclusions—to build knowledge—from the analysis of evidence contained in various historical primary sources, and they must be able to make general statements on large historical issues based on a synthesis of historical evidence and competing interpretations.

In this course, students receive, two days before the exam, three or four essay questions, one of which, they are told, will appear on the midterm. Thus students have an opportunity to organize their responses before the examination period. Similarly, I usually give students a list of 45 to 60 items from which I will select the "Identification" and "Significance" questions. In addition, I give students a considerable degree of choice, asking them to choose, say, four of five or four of six identification questions and, usually, one of two essay questions. In this sample examination, I have provided only the questions that were actually answered.

Your success on this history exam will largely be determined *before* you review for the test in the day or two before the exam. You should begin preparing for the exam starting on the first day of class. Take careful notes in class from the lectures and the discussions. If you lack confidence in your note-taking skills, ask the instructor for suggestions on how to improve. Pay careful attention to class discussions. They often will provide insights into the material, which will assist your understanding. Try to sit near the front of the classroom, and do not miss a class period. Nothing is more important for success in exams than regular class attendance. If you miss a class because of an illness or family emergency, ask the instructor for assistance upon your return.

Read all the required material with care, and engage yourself in the reading at least five days per week. Take notes —whether by underlining, making marginal comments, or writing in your notebook or computer—while you read. Writing ideas down helps greatly in the learning process. As you read a chapter, you may want to underline important points as you proceed. Then review the chapter and write in your notebook or computer the information you think is of greatest importance. Do not underline or take notes excessively. Your job is to cull what is most important. Standard textbooks outline the material for you through their chapter breaks and subdivisions. Let those guide you. If you are not confident that you are getting what is most important from your reading, ask your instructor for an opinion. If reading in this way sounds like it might take a good deal of time, it does. But if you want to succeed on exams, you need to invest time. Indeed, if your schedule permits, it is wise to do some extra reading, perhaps from another history text. In this way you learn different points of view and get a chance to see the material presented in a different way. Devoting regular and sufficient time to study throughout the course will make your review time much more efficient.

Begin to review for a midterm exam at least two days prior to the exam. Cramming the day or two before will not be effective in a history exam. But if you have studied carefully throughout the term, you will be in a good position to review the material effectively. Spending part of your review time with other students is a good idea, but select students who have studied throughout the quarter as carefully as you have.

MIDTERM EXAM

Section A *(50%—24 minutes): Identify and explain the significance of each of the following:*

1. Marcus Garvey
2. sit-down strikes
3. court packing
4. "Rosie the Riveter"

The Identification and Significance questions ask you to write briefly on an important person, event, concept, or issue in American history. The object is not necessarily to write all you know about the item, because that might not be a good use of time, but you do need to address two elements:

1. **Identify the item, answering about it such questions as who? when? where?**
2. **Explain the significance or importance of the item. You might want to think of it this way: After having identified the item, ask yourself: "So what?" Why is this item important to remember? What distinguishes it from being just a trivial fact of American history? To answer the"so what," it is often best to relate the item to other historical issues. Where does it fit in the larger picture?**

ANSWERS

1. Marcus Garvey was a prominent leader of African Americans in the 1920s. His message of self improvement and black nationalism appealed especially to blacks living in northern cities. Much of his work was done through his organization, the Universal Negro Improvement Association, headquartered in Harlem. As part of the black nationalist tradition in African-American thought, Garvey hoped to unite black people in Africa, the U.S. and in Caribbean areas. In some ways, Garvey's movement can be seen as the economic and political counterpart of the artistic and intellectual movement known as the Harlem Renaissance in that they both sought to instill pride in being African American and both sought to revitalize African-American culture. Both drew on the surge of migration of Southern blacks to the north and to the militant attitude of the returning black veterans of World War I. Garvey's influence waned after he was convicted of mail fraud, although he remained a model for later urban black leaders.

It is true that we don't learn much about Marcus Garvey, the man, in this answer. But it is still a good answer because it distinguishes Garvey from others and does an especially good job of

***relating* Garvey to his historical context. Relating the subject of the question to a greater context is really the most important point in answering this type of question.**

A weakness in this response is that one is not sure what the writer means by "uniting various black people."

2. Sit-down Strikes. These were labor strikes in the late 1930s that occurred in the mass production industries like rubber and autos where the workers did not yet have strong recognized unions. The distinctive feature of these strikes was that the workers literally sat down in the factories and refused to work or leave the buildings. Thus, it would be risky for the plant owners to break the strike by attacking or removing the workers because equipment might be damaged. These strikes were important because they led to rapid growth in the mass unions such as the United Automobile Workers, and these unions were part of the newly formed CIO, an organization that helped to organize the semi-skilled in industrial unions and became powerful in the labor movement. The Wagner Act alone had not given workers all the rights they needed.

This answer concisely identifies the item and relates it to the CIO's drive to organize industrial unions. Note that the student literally writes ". . .were important because. . ." This kind of pointing phrase is good because it signals that the student is addressing the "significance" part of the question.

One odd thing about the answer is the last sentence: It is not clear what the Wagner Act is and how it relates to the issue of the sit-downs. It would have been better to leave this out altogether, or, time permitting, to devote another sentence or two showing how the Wagner Act relates to the issue.

3. Court packing is the name given to President Roosevelt's scheme to increase the number of justices on the Supreme Court. Roosevelt wanted to do this in 1937 after his landslide victory in 1936 because the Supreme Court had been hostile to New Deal programs, and he feared it might declare more New Deal laws unconstitutional. Thus he wanted to put on the court more justices who were friendly to New Deal measures. While there was nothing illegal about doing this, it was a very unpopular move because it was seen as obviously political and as insulting to the Court. Even many Democrats friendly to Roosevelt did not favor it. The measure never went through Congress, but it damaged Roosevelt politically. It may have led to a decline in interest in the New Deal and possibly strengthened opposition to it, as soon was exhibited by the growing conservative coalition in Congress. Still, Roosevelt's threat may have been effective because the Supreme Court did approve a number of New Deal laws after the court packing scheme.

This answer indicates nicely why President Franklin D. Roosevelt wanted to "pack" the court. It seems less certain about the effects of the scheme, however. But sometimes even professional historians are unclear about the significance of a historical event. Thus it is good that this student at least *suggested* matters of significance—the conservative coalition, change in the Court's attitude—even if he or she is not entirely certain about these matters.

4. "Rosie the Riveter" "Rosie the Riveter" is a name applied to the women who took part in war work during World War II, especially women who worked in factory jobs that had been held by men. The "Rosie" image was popularized by a famous cover drawing Norman Rockwell did for the magazine *Saturday Evening Post.* The significant thing about this is that women *did* enter the work force in large numbers, more than six million, and in doing so left traditional roles as housewives. Women were applauded for doing this because with so many men in the military and so much war production to be done, it was necessary to fill these jobs with women. And women showed they could do the work well. But women also received criticism from some for working, because of problems on the homefront like juvenile delinquency. Most women wanted to retain their jobs after the war because they paid well and gave them a sense of importance and independence, but most women had to leave them for the returning veterans. Historians debate the long-term effects of the homefront work experience on women, but it does not appear to have changed women's role much, at least in the short term.

A strong answer. A potential problem may be the length of the answer. Do you have enough time to devote this much to a question worth 12.5 percent of the exam? When studying for this kind of question, try to reduce your answer to the essentials. (For example, does the Norman Rockwell detail need to be included?) Time permitting, of course, detail is a good thing.

Section B *(50%—24 minutes): Answer the following question in a well-organized essay. Be sure to support your points with specific historical evidence.*

Discuss ways in which the Great Depression and World War II changed the role of the federal government in American life. In your response, concentrate on how the federal government was used to combat both the Depression and the war.

Like the Section A questions, the essay questions do not require you necessarily to write everything you know about the issues touched on in the question. Rather, you want to concentrate on two things: *thesis* and *organization*. Essays questions will vary in the kind of response they call for. Sometimes they will ask for quite specific information about a narrow topic, or they will list a number of subpoints for you to discuss. However, often, as in this example, the questions are quite broad and call for you to synthesize what you have learned, to interpret the issue—to bring meaning out of the evidence of the American history you have been studying.

The thesis is a statement that indicates the argument of your essay, the answer, in brief, that you offer for the question asked. Argument is the outcome of your thinking process, a series of connected statements or reasons that establishes a position you want someone to accept. Upon seeing an "essay question," ask yourself: "What thesis best explains this historical issue?" The thesis suggests to you and to the reader the direction your essay will follow, and therefore should be stated in an introductory paragraph. Think of the rest of the essay as the place to present evidence and to develop your main ideas. The topics and information contained in your essay should relate in some way to the thesis. Include enough specific evidence (facts) to convince the reader that the thesis is a valid one. You might also need to explain why contrary evidence does not invalidate your thesis.

Time is limited: Again, your task is not to say everything you know about the topic in question but to explicate your thesis, a thesis you have selected because you think it is the most significant one pointed to by the question.

For proper organization, briefly *outline* your essay before beginning to write. If you do not, you will be listing facts rather than writing a historical essay. Before writing consider the following steps:

- Study the question and arrive at a thesis that you think is best
- Determine the principal issues you think are necessary to support the thesis
- Arrange your main ideas (planned paragraphs) in a logical order (outline) for the development of your thesis.

An effective introduction might also tell the reader the main points you will be discussing in addition to the thesis. A conclusion, if time permits, should not simply repeat material in the body of the essay, but should offer fresh comments to summarize your answer.

Basic to the organization of your essay is the paragraph. At least one paragraph should be devoted to each of your main points. Keep the paragraphs independent and coherent. Begin a paragraph with a topic sentence. You then proceed to give evidence that supports the topic sentence and thus your thesis. You may want to write more than one paragraph on a given point. Just keep in mind that you should start a new paragraph when beginning the discussion of

another main point paragraph with a topic sentence. You then proceed to give evidence that supports the topic sentence and thus your thesis. You may want to write more than one paragraph on a given point. Just keep in mind that you should start a new paragraph when beginning the discussion of another main point.

ANSWER

In the late nineteenth century and during the Progressive era of the early twentieth century, the federal government grew in numbers of employees and agencies, and began to take on some regulatory and welfare functions for the nation. But the achievement of a more substantial welfare state and a central position in the life of the nation occurred for the federal government in the period from 1930 to 1945 largely in response to two major crises: the Great Depression and World War II.

The introduction is good in that it provides some background and states a theme of development of a welfare state and a central position for the federal government. Notice also that it does not repeat the question. Too many students waste time by simply repeating the question in the introductory paragraph.

The Great Depression set in after the stock market crash of 1929 and was so severe that by the time of Franklin Roosevelt's inauguration in 1933, the gross national product had declined by about half, huge numbers of businesses and banks had failed, and about 25% of workers were unemployed. Herbert Hoover had not been inactive in fighting the Depression in its early years, and this "forgotten progressive" did more using the federal government than is commonly thought. Nevertheless, his steps were hesitant and timid, considering the magnitude of the problem. Roosevelt and the Democrats could understandably take their victory in the election of 1932 as a mandate for exercising greater government strength in combating the depression. And this they did. In the famous first "hundred days" of the administration, numerous measures were passed by Congress to fight the Depression, and these largely involved the greater exercise of power by the federal government. For example, the National Recovery Act and Agricultural Administration Act, cornerstones of the early New Deal, involved attempts at unprecedented government planning and management of the economy while also involving private groups.

This is not to say the governmental structure of the country radically changed. Although the TVA was a considerable success, New Deal measures did not lead to a system of federal government planning for the economy. NRA had severe weaknesses and was declared unconstitutional by the Supreme Court. Socialism was not part of the array of New Deal programs and it won little popularity at the polls. Rather, the New Deal expanded the power of the federal government by involving it in more areas of society, first in an attempt to control production, with NRA and AAA, and then later with attempts to stimulate demand. For example, the Wagner Act gave greater strength to laboring people in their efforts to organize; the WPA made the federal government an employer of last resort, the Social Security Act provided the framework for a limited welfare state, and the Fair Labor Standards Act set minimums for wages and hours. All of this helped create what today is called the "social safety net," a bare minimum of material support for citizens. By the late 1930s the New Deal had pretty much reached its limit when a coalition of Republicans and conservative Democrats blocked further legislation.

This question lends itself to a fairly simple organizational structure, and this essay has done it well. After an introduction, it moves easily into the Great Depression years, with brief but sufficient detail on the extent of the crisis. (No more is needed on the Depression, because the question is not about it.) The essay does not describe in much detail what the New Deal measures actually did, but it does give a sense of the activist Roosevelt administrations, which turned people's attention more toward Washington, D.C. Then the response moves to the war.

World War II, with the United States as the "Arsenal of Democracy," put an end quickly to the remnants of the Great Depression because of the great need for war production. Unemployment virtually disappeared. The statistics on wartime government spending and production that show an end to the Depression also argue for an increasing role for the federal government in American life. The federal government rationed consumer products and, in some cases, as in autos, ordered an end to their production. For the first time, the government began withholding federal income taxes. The government, through various boards, allocated scarce raw materials among competing industries. It mobilized women into the work force and then persuaded them to leave it. It gave generous benefits to veterans. It drafted large numbers of men into the armed forces. It influenced the tone of movies. It took steps, albeit timid, to end discrimination by race in war industries.

By 1945, the American people had a far greater familiarity with and reliance on their federal government than they did in 1930. That it took an economic emergency and the greatest war in history to bring the federal government to a position of great power shows that this outcome was not necessarily the wish of the American people, but something brought on by external circumstances. Management of a modern economy and a leading role in world affairs rendered the national government of the nineteenth century obsolete.

Nothing is really said about the war itself, but that is appropriate, because, again, the question deals with the growth of the federal government and its role in fighting the war, rather than the war itself. Specific examples are given of new roles for the federal government. The conclusion is excellent. It does not repeat what has been said but, rather, draws the reader to larger issues.

Some weaknesses to note: Although Hoover is properly mentioned, no specific detail is. That he used the federal government is simply stated, rather than demonstrated. Similarly, when statistics regarding World War II are mentioned, no specific examples are given.

This essay response is a good example of a successful strategy. From the beginning, the student takes care to answer the question being asked. Too often, students get off the track and start writing about events and issues not pertinent to the question. Writing about extraneous issues, even if the information is correct and the ideas interesting, will detract from the grade one receives, often seriously so. A grader must judge all exams in the same way and cannot give credit for answers—however well stated—that are unresponsive to the question being asked. Sometimes students will get off the track intentionally, because they really don't know how to answer the question. They decide, in effect, to answer another question and try to fool the grader. Few graders are fools. But even students who study hard and know the material can also stray if close attention is not paid to the question.

It is a good idea to *sketch* an outline of your response before you begin to write. This will help to focus the response, ensuring that no major points are left out and that no extraneous points are included.

This question, like many broad essay questions, asks you to "discuss." It's important to realize that "to discuss" is not the same as "to describe." Discussion includes a certain amount of description, to be sure, but discussion also involves examining, investigating, arguing, and debating. When you discuss something, you take a position and support that position with evidence.

FINAL EXAM

The final exam counts for 30 to 40 percent of the course grade. Five to seven days before the exam, students receive four or five essay questions, two of which, they are told, will appear on the final. Students are also given a list of 45 to 60 items from which the Identification and Significance questions are selected. As with the midterm, students have a choice of answering four of five or four of six questions from Section A and two of three of the essay questions in Section B.

Section A *(20%—22 minutes): Identify and explain the significance of the following:*

1. The Stamp Act
2. The Middle Passage
3. War of 1812
4. Dred Scott

ANSWERS

1. The Stamp Act was a revenue-raising measure passed by the British in 1765 requiring American colonists to pay a tax for stamps that were put on a variety of paper products and documents. It followed the Sugar Act of 1764 as the British government attempted to raise revenue for the defense and administration of the American colonies. The colonists, however, saw it as an unconstitutional law—"Taxation without Representation Is Tyranny"— and there was strong resistance to the measure, including mob violence in the city of Boston. This act led the colonists to consider deeply their constitutional relationship to England and begin a resistance to British measures that would not end until independence was declared.

This is a good answer because, by giving the date and a brief explanation of the act, the student has distinguished this measure from others passed in the time period. Certainly more could have been said to describe the act, and perhaps it could have been said better, but what we have here is sufficient for identification. The response also answers the issue of significance by placing the Stamp Act in the context of the American revolution and by showing that the colonists protested it strenuously on constitutional grounds.

2. The Middle Passage was that part of the slave trade dealing with the voyage from the African slave markets to markets in the Western Hemisphere. Africans were packed onto congested ships in very unhealthy conditions. A large percentage typically died on these voyages from malnutrition, disease, and suicide. Many of the ship's crew also died. The slave trade fed the great demand for labor in the western hemisphere and contributed to the establishment of the "peculiar institution" in the American South. Slavery would be a key element in the eventual split leading to the Civil War.

The description doesn't provide much detail about the voyages and the horrors on board ship. Nevertheless, the answer concisely says what the Middle Passage was. It also ties it to the larger issue of slavery and the Civil War. A rather general answer, but still a good, concise one. It's important to keep in mind that time is limited and not everything can be said.

3. The War of 1812 originated with the Americans disputing neutral rights with the British and French during the later part of the Napoleonic Wars. Although the United States had grievances against both countries, war was declared against the British, who had a greater navy and thus were in a position to do more harm to American shipping. Also, relations with England had never been good since the revolution. Except for some naval victories (especially Commodore Perry's on Lake Erie), the war was basically a disaster for American forces until the final battle at New Orleans. An invasion of Canada failed, and the British burned Washington, D.C., and bombarded the fort in Baltimore harbor, where Francis Scott Key wrote the *Star Spangled Banner*, which was based on an English drinking song. The Battle of New Orleans actually took place in early 1815 after peace was settled (slow communication meant Americans did not know this). This battle was a decisive victory for the U.S.—more than a thousand British casualties and hardly any American. It instilled a perhaps unwarranted sense of pride in American arms and made a hero of Andrew Jackson. The war is sometimes called the Second War for Independence because relations with England improved somewhat after the war and the U.S. received more respect from European nations.

This answer is good because it contains responses to the two basic elements of this type of question. It identifies the war and it explains some of its effects or importance. One wonders, though, whether it could not have been answered more concisely. (Think of ways this could have been done to save time for the rest of the exam: for example, is it necessary that the *Star Spangled Banner* information be included?)

4. Dred Scott was a slave whose master took him to live for a while in a free state and a free territory before returning him to a slave state. Scott sued for his freedom based on his residence in the free territory. The case went to the Supreme Court, where it was decided in 1857 with Chief Justice Roger Taney writing the main opinion. Taney argued that African Americans, even Free Blacks, were not citizens and thus that Scott could not sue. But he went beyond this to argue that the Missouri Compromise was unconstitutional. Thus Congress had no authority to regulate slavery and many years of compromise over the issue were negated. While the South was overjoyed, this decision inflamed passions among Republicans, bringing into question the party's reason for being, and making it harder to compromise on the growing sectional division within the U.S.

The identification part of this answer is not detailed. For example, we don't learn what state or territory Scott lived in; however, the response does provide the essential background for the case, the elements of the decision, and its importance for the times. It is not necessary in these questions to get bogged down in too much detail.

Section B *(80%—86 minutes): Write well-organized essays in response to the following questions. Be sure to support your points with specific historical evidence.*

1. Within the seventy-five years following the Declaration of Independence, the United States expanded from the eastern seaboard to the Pacific Ocean. Discuss the nation's acquisition of territory from 1783 to 1850, with specific attention to reasons for expansion, methods used to acquire territory, and obstacles to expansion.

2. In 1861 the United States became disunited. Discuss how this happened by comparing and the "North" and the "South" in the evolution of American society from 1607 to 1860.

Remember, without an effective structure for your essay—a thesis and good organization—your answer will be little more than a motley compilation of facts, having little focus, and hence little meaning. But it is also true that the intelligent determination of a sound thesis and effective organization is not possible unless you have a good understanding of the evidence (facts) and the context of American history.

Two extremes of error are typical in responses to essay questions. One type of error is that a student has learned the detail, the facts, very well, and they come pouring out in an essay that lacks a thesis and is without ideas to connect these facts. What you have is a listing of much detail, which may very well be correct and might show that you have been studying and memorizing, but no statement or interpretation to help in understanding the large historical issue. This is a failure to synthesize.

The other type of error is the opposite extreme. The student's essay contains a number of

good ideas, perhaps a good thesis, supported by a number of important subpoints, but there is little or no specific evidence given to support the larger points. It's possible in this case that the student hasn't learned the material in any depth, but has paid enough attention to pick up on some main points. More likely, the student who makes this kind of response has analyzed the evidence but fails to write a proper essay, fails to understand that one's ideas cannot be accepted on faith, that they must be supported by some specific evidence.

ANSWERS

Essay 1

The American people were an imperialistic people with their own continental empire in the period up to 1850. That the United States used a variety of methods to expand from sea to sea in less than 75 years shows that the American people were bold and aggressive, bent on expanding their civilization despite a number of obstacles.

The first expanse of land added to the original thirteen states came through the Peace of Paris ending the Revolutionary War in 1783. To the surprise of most Americans, England relinquished the land west of the Appalachian mountains to the Mississippi River and south of Canada, more than doubling the size of the new republic. The Northwest Ordinance of 1787 set up the structure for territorial government and establishment of statehood in the new lands. This important measure helped determine that the American continental empire would be unlike traditional empires: new territory would eventually be brought into the union with the same rights and privileges as the original states.

Purchase brought in the next major pieces of territory. In 1803 the U.S. bought the Louisiana territory from France because President Jefferson and others wanted the port of New Orleans. Despite constitutional scruples, Jefferson took advantage of the offer of all Louisiana, and he soon sent Lewis and Clark to explore it. Another acquisition came with the purchase of Florida from Spain in 1819 engineered by the expansionist secretary of State, John Quincy Adams. Encroachments upon Florida territory by Americans had led Spain to believe they must sell because Florida was no longer defensible.

Warfare brought in additional territories. Many Americans went into the Texas territory of Mexico in the 1820s and 1830s and others helped support the Texas rebellion against Mexico in 1836. In 1845 Texas was annexed to the U.S. The next year a war with Mexico started that should be seen as an aggressive war of American expansion. President Polk was especially interested in acquiring the port of San Francisco, and he used a military incident on the American-Mexican border to start a war, which American forces won decisively. The treaty in 1848, part of which involved payment from the U.S. to Mexico, brought a vast territory in the Southwest and California to the United States. That same year, after threats of war, a treaty between the U.S. and England divided the Oregon Territory.

The rapidity with which this territory was acquired shows how aggressive and land hungry Americans were. Some Americans wanted all of Mexico, others wanted Cuba, some wanted Canada. The land hunger was in part based on economic concerns; the country was after all primarily agricultural and land remained the main source of wealth. It was also in part strategic as the U.S. vied with other European nations for primacy in North America. But it was also ideological. "Manifest Destiny," a term first used in 1845 by John L. O'Sullivan, a newspaper editor, was a vague but powerful idea that seemed to drive Americans or at least was a way for Americans to justify their actions. Basically it was a belief that God, Providence, or whatever meant for Americans to take over the continent and extend their institutions. To Americans it meant extending "the blessings of liberty" to other less fortunate and less civilized peoples.

Obstacles to empire included geographic challenges such as the many rivers to ford, mountains to traverse, harsh climate, and the "Great American Desert"—the Great Plains thought inhospitable to settlement—but the less "civilized" people presented the most difficult problems. War and treaty overcame opposition from foreign peoples, principally the Mexicans and British. But the original inhabitants of the land, the Native Americans, remained an obstacle. Indian-white relations show examples of humanitarianism and there was a strain of thought among whites that Indians were dignified "noble savages," but the overriding theme is one of bloody conflict, starting as early as the founding of Jamestown. The Indians in the lands ceded by the British in 1783 were driven out by the late 1830s through warfare and the Indian Removal Policy of the Jackson administration. In the trans-Mississippi West, warfare proved relentless in driving the Indians off land thought desirable to whites. Indians of all kinds were the chief victims of Manifest Destiny, for little room could be found for them within the American definition of civilization.

In conclusion, the westward expansion was a folk movement, driven principally by the individual decisions of hundreds of thousands of aggressive people who sought opportunity, if not riches, in the American West, and who believed they represented the best of human civilization. The U.S. government did not plan or guide this expansion, though it did sponsor exploring expeditions and conducted war when necessary to ease the way. American expansion from sea to sea was one of the most effective imperialistic adventures in history.

The question asks one to address several subpoints, but doesn't give much of a clue to a thesis. In the introductory paragraph, we are told that Americans are aggressive and imperialistic, and this theme is carried well throughout the essay.

There are a number of ways to organize a response to this question. The writer in this case has chosen to describe the acquisition of territory while also explaining methods used, and then proceeds to deal with the two remaining subpoints, reasons and obstacles. This structure works satisfactorily. But perhaps you can think of a more effective way to organize the essay.

A danger in broad questions like these is that answers often lack specificity. This response is good in that it is specific about chronology. It ignores some treaties remedying boundary disputes with England, but it certainly touches on the main points of expansion. Frequently, we would have liked more detail, perhaps more about the Mexican War and settlement of the Oregon territory, but, in fairness, the writer's time *is* limited. Not everything can be said, nor should one attempt to do so. While maybe not as detailed as we might like, this response is reasonably specific, and no major items have been left out.

Essay 2

It is tempting to say that the Civil War was inevitable, given the strong regional differences that are evident from early colonial days in the territory that became the United States. It was not, in fact, inevitable; but the regional differences did contribute to a sense of distinct identity that, combined with controversy over the expansion of slavery, led to the failure of compromise within American politics and thus to secession and war.

This question is somewhat tricky, because it implies that regional differences caused the Civil War. The writer has done a good job in the introductory paragraph of dealing with the issue of a theme. We learn that the response will be dealing with the development of regional differences, as the question seems to want, but it also states clearly that other elements were involved in ultimate disunity.

The English colonies that were the nuclei for later American states all have their unique histories. They were founded for different reasons, by people with different cultures, and they had different geographic conditions. One can think of colonial America as having various regions, say the upper and lower South, the Mid-Atlantic, and New England. The Virginia and Massachusetts colonies can serve as examples of regional differences. Massachusetts was settled by Puritans with a specific mission in mind, the "City upon a Hill" ideal by which they hoped to establish a community that would be a model for England and the world. In so doing, they ran their colony with close supervision and strict adherence to religious principles, building a "new England" of nucleated villages and diversified subsistence farms. In Virginia and the South, the soil and climate lent itself more to the growing of staple crops, especially tobacco, along with rice and indigo, and, later, cotton. Virginia was founded by a profit-seeking company that was less organized, less religiously imbued, and less governed by ideology than Massachusetts. Because of the nature of its geography and the ability to grow staple crops, its settlement patterns were different, with population more widely dispersed, few towns, and a large slave population.

The preceding paragraph demonstrates that not all the detail one might know can be easily crammed into an essay constrained by time. The paragraph does seem to provide enough information to show that regional differences existed from earliest settlement. Perhaps some other details would have done a better job of illustrating this point.

It should be noted that, during the American Revolution, revolutionary leaders came from various regions of the thirteen colonies. Both southern and northern figures were among the leaders. Both northerners and southerners strongly believed in the republican nature of the revolution. Still, North-South differences were among the differences noted at the Constitutional Convention in 1787. And at the heart of the differences was the issue of slavery. Slavery, never as important in the North, now was being gradually eliminated there, but emancipation gained little headway in the South. Compromise was the order of the day at the convention to keep the sides united.

The preceding paragraph is brief but accomplishes matters of importance: It indicates the regions had things in common, too, thus not exaggerating regional differences. It also introduces slavery as a point of divisiveness—something the introductory paragraph said was important.

Slavery continued to bedevil Americans in the nineteenth century, but another major transformation was also occurring to distinguish the two regions. Great changes in the economy, recognized now as a market revolution and the industrial revolution, took root much more deeply in the North than in the South, such that, by 1860, the North outstripped the South in virtually all measures of an industrial economy, including population and the extent of urbanization. Immigrants noticed the differences, and, in the great immigrations of the antebellum years, they overwhelmingly settled in the North. The economic importance of the South, with King Cotton the major factor in American international trade, should not be underestimated, but the point is that the two regions were developing quite differently.

These remarks on economic change make a point of considerable importance. It would be strengthened if more detail, some statistics, for example, were given to show differences between the regions.

The other great development was the westward expansion of the nation. Both regions thought expansion a good thing, and people from both areas took part in it, but as they moved west they also brought with them their different cultures, and expansion eventually led to a split in the nation over the issue of bringing slaves into the territories. Southerners feared that the pattern of their declining political power in the federal government would eventually mean the extinction of slavery and the particular society that its leaders enjoyed.

This paragraph on expansion also discusses an issue of major importance. It would be improved, however, if specific mention were made of some of the events and compromises concerning slavery that accompanied the expansion of the nation, for example, the Missouri Compromise, Compromise of 1850, and the Kansas-Nebraska Act.

It's easy to overstate differences in the regions. They shared many important things: language, revolutionary traditions, pride in expansion, among others. But differences in regional development, with the existence of slavery in the South by far the most important, created in each side an individual identity, and, in the case of the South, an identity sufficiently strong to contemplate and then implement steps to separate into a distinct nation.

The conclusion is good because it reminds the reader not to exaggerate regional differences, but states with confidence that the phenomenon of regional identity was central to disunity, and this, after all, was the main topic of the question.

CHAPTER 23

PURDUE UNIVERSITY

HISTORY 151: AMERICAN HISTORY THROUGH 1877

Aram Goudsouzian, Graduate Teaching Assistant

TWO EXAMS ARE GIVEN IN THE COURSE, A MIDTERM AND A FINAL. BOTH EXAMS ARE strictly essay exams.

This is an introductory survey course that covers colonization through Reconstruction. For both the midterm and final, students are given four questions and are expected to write on three of them. Each essay is worth 33 points. (They get one point for signing their name.) They should spend roughly an equal amount of time on each question, and, since they have two hours, should have no trouble completing the exam in time.

The course is designed to give students a familiarity with basic trends in American history, but also to introduce them to how historians approach their craft. If students understand the subjective nature of history, they are more willing to challenge traditionally held assumptions, to engage in debates over historical interpretations, and to create convincing arguments in written work based on available evidence.

MIDTERM EXAM

Write complete, well-organized, and clear essays for three of the following four questions. You have two hours to complete the exam.

1. **The Virginia and Massachusetts Bay colonies present a study in contrasts. Why did Virginia experience so many problems in the first several decades of the**

settlement yet eventually flourish, while Massachusetts Bay succeeded so dramatically in its early years yet experienced a social and cultural disintegration toward the end of the 1600s?

ANSWERS

The Virginia and Massachusetts Bay colonies' experiences certainly did contrast in the 1600s. Massachusetts Bay was more successful in its early years because the Puritan settlers established order and hierarchy, unlike the Virginians. Yet part of the reason why Virginia eventually flourished was because they allowed for private property and the introduction of capitalist competition. Massachusetts's Puritan society could not allow for such developments, and disintegrated in the face of them.

Since the exams are entirely essays, I stress the importance of organizing ideas ahead of time. Although many students are averse to them, outlines are the most effective way to prepare for essay exams. A study guide passed out the week before the exam should give students a better chance to anticipate the main concepts that will be stressed in the exam.

Virginia was originally settled by the second and third sons of wealthy English gentry. They desired the land, status, and commercial prosperity that was unavailable to them in England. However, because they were gentry, they were unwilling to work hard, and early colonies like Jamestown experienced widespread starvation and death. Also, these early colonists had problems with native Indians, and often used violence as a solution even when Indians tried to cooperate. Finally, they chose a site for colonization that was swampy and malaria-infested.

English stockholders were unhappy with the problems and lack of economic development in Virginia and instituted a number of reforms under Sir Edwin Sandys in 1618. Sandys allowed for private ownership of land rather than the original communal system. He also instituted a headright system, by which additional land grants were awarded to settlers as they brought more people to the New World, and set up local governments to make the colony more orderly. These reforms helped solve the labor shortages and lack of order in early Jamestown.

As tobacco became more popular in England, Virginia truly flourished. Yet tobacco was a labor-intensive crop and needed a large body of workers.

Indentured servants originally filled this need, but they were generally free after 4-7 years of service and often became competitors in the tobacco economy once free. Tobacco planters increasingly turned to the use of African slaves, especially as life expectancies rose and owning a slave for life became a better investment than an indentured servant. A system that codified racial behavior was implemented to promote social order and economic development, to the obvious tragic detriment of slaves.

The student fully develops the situation in one colony before going on to the other. This is a good strategy and creates a cogent essay. The only danger is the possibility of failing to apportion limited time adequately. Be certain to keep an eye on the clock, to ensure that you'll have sufficient time to develop the other half of your case.

Massachusetts Bay, on the other hand, did not experience these same early problems. Massachusetts was originally settled in 1630 by Puritans, a radical Protestant reform group in England. Puritans believed in predestination, the idea that one's salvation was determined before they were born. This belief actually contributed to social order, as people believed that if they were saved, it would manifest itself in how they lived on earth. Puritan society was also rigidly hierarchical, and church leaders generally had access to power that other citizens did not. This type of order was successful when establishing the colony because there was a distinct group of leaders, and these leaders generally served the needs of the loyal and faithful. The system worked, as stable settlements flourished, life expectancy rates were high, and there were few poor people.

Note the phrase "on the other hand" in the first sentence. Always be careful to signal to the reader any change in subject, argument, or point of view. Phrases such as this are valuable sign posts that make it much easier for your reader to follow the course of your argument. In an essay exam, your grade will benefit by your effort to provide a clear, straightforward reading experience for the examiner.

Puritan society disintegrated because its leaders could not maintain this system of rigid order in the face of private ambition and lack of faith among succeeding generations. The Halfway Covenant in 1662 was an attempt by leaders to compromise and allow for a larger body of church members. Property disputes and the emergence of a prosperous new merchant class outside of the Puritan elite signaled the new problems in once-harmonious Massachusetts. The Salem

Witch Trials in 1692 are probably the most dramatic example of the social and economic tensions that developed, as more traditional, rural Puritans accused those who were more a part of the commercial economy that they were witches.

This essay is well organized, with a clear argument in the first paragraph. It then chronologically traces the development of the Virginia colony, with a paragraph on its problems, a paragraph on reforms, and a paragraph on reasons why the colony flourished, it then spends two paragraphs on Massachusetts, beginning with its original success and then detailing the disintegration of Puritan society.

The essay makes good use of historical detail, using specifics like the headright system and the Halfway Covenant to illustrate larger ideas within each paragraph. The essay is strongest, however, for the first paragraph, which grasps the larger concept of how both order and ambition affected each colony's success.

2. Prior to 1763, American colonists experienced little direct "oppression" at the hands of the British government. Yet by this time a number of economic, political, and social forces conspired to make a revolution of thirteen united colonies possible. What were these forces, and how did they influence how the colonists saw both themselves and the British government?

ANSWER

While very few Americans viewed independence as a viable or desirable option prior to the 1760s, a number of factors made colonists willing to see themselves as both different from England and united as thirteen colonies.

Note the qualifying statement beginning this response. Such qualifications may be necessary to avoid a charge of making blanket generalizations. Use them wisely.

For one, American colonies never established clear lines of authority and order with England. England had a number of both internal and international conflicts throughout the seventeenth and eighteenth centuries, and sometimes focused little attention on the colonies in the New World. Without strongly established legitimate authority, conflicts periodically erupted over who deserved to rule in the colonies. For instance, Bacon's Rebellion in 1676, in which Nathaniel Bacon challenged the authority of Virginia's House of Burgesses and established an outlaw militia group, illustrated that tension.

Similarly, James II's establishment of the Dominion of New England, which wiped out local self-government in the northernmost colonies in 1685, was met with public outrage. When James II was forced to abandon authority, there was a brief period in the 1690s when there was no clear system of government in Massachusetts, New York, and some other colonies.

At the same time, the thirteen American colonies increasingly began to see themselves as united in purpose, as economic growth and increased mobility created lines of communication between the colonies. The rate of population growth skyrocketed in the early 1700s, and people spread throughout the colonies. Since many colonists owned land and enjoyed economic prosperity, they began to trade within the colonies. There were more consumers and higher incomes. The colonies also banded together as part of the British conflict with France during the 1700s, which often spilled into the New World. For instance, the French and Indian War from 1756 to 1763 spanned the entire colonial frontier.

As the body of American consumers grew, British merchants and bureaucrats saw the profit potential of the New World not only as a source of raw material but as a potential source of taxation and a place to sell goods. American merchants and planters began to find themselves increasingly undersold by British goods. Also, the British government began to enforce more of the regulations against smuggling and corruption, which had always existed but were previously rarely enforced. The merchants and planters who were hit hardest by the increased British presence were the ones who would lead the struggle in the American Revolution.

This question can be difficult because it requires the student to bring in information from a variety of lectures and angles of analysis. A good essay can bring them all together. Here, the opening sentence frames the rest of the essay, which is organized into three paragraphs dealing with questions of legitimacy, sources of colonial unity, and reasons for British resentment, respectively. The student should understand the tenuous hold of the British government on the daily lives of American colonists, although they were loyal British subjects. Also, it's important for students to realize that the push against British regulation came specifically from a certain class within American society.

3. To what extent was the American Revolution actually a revolution? Was the American Revolution originally fought to protect all people's interests? Did it change the social fabric of the country or alter the way Americans viewed government?

Always read the question carefully. Note that this is a multiple-question question. Be certain that you answer each of the constituent questions. In fact, you can usually use the several questions as the basis of your essay's outline. Answer them in order of the asking. They are an aid to organizing your response.

ANSWER

The American Revolution overthrew the existing British government, but, unlike traditional revolutions, was not originally motivated by a desire to put a radically new government structure in place. However, the Revolution was justified by republican ideals, which changed Americans' conceptions of the role of government. In this respect, the American Revolution was indeed a revolution.

This essay, like the others, has a clear first paragraph that outlines the rest of the essay. This first paragraph is by far the most important in any essay, because it creates a framework for the argument and displays to what extent the student has grasped the larger conceptual issues at stake.

The displeasure with the existing British government stemmed originally from the class of merchants and planters who were most negatively affected by new policies of increased taxation and regulation in the mid-1700s. This business class had for years depended on British protection in international conflicts, yet had been left to itself for the most part in matters of trade. Colonists regularly sold goods to merchants from countries besides England, and smuggled in goods as well.

By the 1760s, the British government realized the profit potential in the New World, and King George III led a reform movement that involved a more active role in colonial affairs. With Parliament, George enacted laws like the Sugar Act, which created increased regulation against smuggling, and the Stamp Act, which was the first direct tax on the colonies.

The merchant class, most hurt by these policies, led the cries of indignation against the King's authority. These men were fighting not for a widespread reorganization of society, but to preserve the economic and social privileges that had been theirs for most of the century. Although the King was exercising royal prerogative and in fact trying to reduce corruption, these American radicals laid charges of corruption and tyranny against the King. These well-organized radicals led a propaganda campaign painting British actions as infringements of traditional liberties. For example, radicals characterized the Boston Massacre as the work of cruel British troops who fired into a crowd with little agitation.

This answer diverges from the popular view of the Revolution as a *popular* revolution. Generally, the student supports this view well; however, note that the last sentence in the paragraph above leaves a loose end. The student implies that the popular portrayal of the Boston incident as a "massacre" was mere propaganda. This assertion needs support.

For an overthrow of British authority to be successful, the radicals had to convince the majority of the public to follow their leadership. Influenced by Enlightenment thought, these radicals suggested that the King could not rule without the consent of the people. The ideals of republicanism—equality, liberty, virtue—became the hallmarks on which the majority of Americans framed their sympathies for the Revolutionary cause. In the Declaration of Independence, Thomas Jefferson wrote that "We hold these truths to be self-evident—that all men are created equal." This was a radical departure from previous political thought, which had allowed for the authority of the king based on divine right.

The American Revolution was a revolution in how it changed how Americans believed a government should be run. Without this commitment to republican ideals, the revolutionary struggle would have remained in the hands of a small minority of merchants and planters. Yet the nation was founded on new principles that affected what Americans considered just and legitimate.

To answer the question of whether the war was actually a revolutionary movement, one must address the historical background. A good answer will not just answer yes or no to this classic question, but will incorporate in what respects it was and was not a departure from previous norms.

A good answer in effect explains both the causes and effects of the Revolution. Students must break from the traditional notion of British oppression of all the people, and understand that revolutionary sentiment came from a particular class with economic self-interests. If the student understands this, the next logical step is to see that this does not constitute a revolution in itself.

Yet understanding what does make this a revolution is perhaps the central concept of the first half of an early American history survey course, because it creates the standards and ideals on which Americans base their actions to this day. To achieve the consent of most people, a new government would have to incorporate republican ideals.

4. What was the design and intention of the 1787 United States Constitution? Why was there a bitter public debate over its ratification? How did this debate

fuel the political conflicts of the 1790s and the early 1800s?

ANSWER

The Constitution of 1787 was put into place to centralize authority at the national, rather than the state, level. It was designed and defended by Federalists, who believed that a national government best represented the long-term interests of the nation. They promoted policies designed to stimulate a large central government and economic development. The original opponents of the Constitution were known as the Anti-Federalists and believed that power should rest at the state level to best represent the interests of the people. The Republican Party represented these ideals after the ratification of the Constitution.

The thirteen states, after declaring independence, originally agreed upon a loose national structure under the Articles of Confederation. The Articles reflected the states'self-image; each state saw itself as distinct, with different interests and ideals. This idea was reflected in the variety of state constitutions that were drafted upon independence, from Pennsylvania's democratic single house of legislature to Maryland's elitist structure, which limited political power to a small, land-owning minority. There were problems under the Articles of Confederation, including unstable state currencies and an inability to pay national debts.

The student refers to "thirteen states" rather than "thirteen colonies"—a mistake, but one that most instructors would forgive, ascribing it to the pressure of a time-limited exam. Nevertheless, try to reserve time to read over your response in order to correct such careless errors.

Nationalist leaders, rather than reform the Articles of Confederation, decided to scrap them and create a new constitution. Most of these men represented economic interests which would benefit from a uniform national trade policy and a government which encouraged economic development. Although compromises were made to the interests of popular political representation, such as the House of Representatives, the United States Constitution allowed for a much stronger role for the national government, including the power to tax, print a national currency, regulate interstate commercial disputes, and provide a united front for international issues of trade and diplomacy.

State conventions for ratifying the Constitution involved often bitter disputes. Opponents argued that a national government would usurp power from the people, since their state representatives most represented local, common interests. Federalists countered with a defense that included the Federalist Papers, a series of written arguments by James Madison, Alexander Hamilton, and John Jay that argued that the Constitution was necessary and that a federal republic would actually best benefit the American public. Eleven of the thirteen states did eventually ratify the Constitution, although often by close margins.

The Federalists, under the presidency of George Washington and John Adams, dominated American politics until 1800. Led by the influential Secretary of the Treasury Alexander Hamilton, the Federalists created a national debt by assuming state debts, set up a national Bank of the United States, established a strong national defense, and passed federal taxes. These policies were designed to promote a loyalty to the federal government among the people and to establish a large bureaucratic structure that could promote national interests.

The Republicans under Thomas Jefferson came into power in 1800. Bolstered by fears of common people that too much central authority would result in tyranny similar to that of England, the Republicans claimed they represented the interests of the people. Jefferson led a movement to scale down the size of the national government, cutting back federal offices and the military. Jefferson believed that power at the state level more accurately reflected what common people needed. Although Jefferson approved policies that contradicted his ideology, such as the Louisiana Purchase and the 1807 trade embargo, his presidency represented a shift away from Federalist excesses and reflected a fear of too much central authority.

The same tensions and issues that dictated the struggle over the ratification of the Constitution survived into the following decades, as Federalists and Republicans debated over at what level power should be located.

The key question in this essay revolves around who should rule to protect the interests of the American people. One must first understand that the Constitution was not an inevitable political result, and that it represented the interests of only some of the people. The student must understand why some would support the Constitution and centralized authority, while others would find fault with it.

Once that is established, the student should be able to draw a connection between that political conflict and the one that ensued around the turn of the century; they were battles fought on similar principles. There are numerous policies among the Federalists and Republicans one can draw from, and only a few are mentioned here.

As with all historical essays, clarity is the main objective. Each paragraph makes one key point that is illustrated with examples, and the paragraphs follow a logical flow.

FINAL EXAM

The key to general preparation for essay exams is to organize your thoughts in outline form. Making outlines forces students to think in terms of a logical flow to an essay. Each essay question has a "big picture" concept to it, which students must recognize. If they grasp the main point of each essay, then they can tailor their outlines to illustrate that point, and study details to provide evidence of that point.

Write complete, well-organized, and clear essays for three of the following four questions. You have two hours to complete the exam.

1. Discuss some of the major changes in the American economy in the first half of the 1800s. What was the relationship between these changes and some aspects of the reform movement (both religious and secular) that flourished during this era?

I first expect students to develop writing skills—not necessarily to write more eloquently, but to write with clarity and organization. This writing skill *will* benefit them no matter what profession they choose. Second, I expect students to develop an ability to express themselves orally. I find many students generally are uncomfortable making assertions in the classroom. Participation and debate in class discussions makes students learn more actively, rather than passively accepting information and churning it back out on the exam.

ANSWER

The United States underwent broad, significant economic change in the first half of the nineteenth century. While these changes brought economic development, they also transformed traditional social and family relationships and left many in search of new means of achieving order in their lives. Reform movements answered part of this need for community, social order, and benevolence that was being ignored by the new capitalist system.

The opening paragraph is the most important part of any essay response. It may be helpful to adopt the "inverted pyramid" model long familiar to journalists. Construct an opening paragraph that contains the essence of the response, so that it could stand alone, if necessary, and still serve as an answer (not a full answer, but an answer) to the question. The remaining paragraphs flow from the opening paragraph, amplifying and defining it. If you write an "A-level" first paragraph, the chances are reasonably good that the instructor will be inclined to judge the entire essay as A-level work.

At the turn of the century, the country was overwhelmingly rural, and cities were relatively small. Most forms of production did not occur in factories or urban areas; rather, households were the most significant place for production. These methods were transformed as canals and railroads connected more of the American economy, agriculture specialized in a single crop on a large scale (like cotton in the South and wheat in the Midwest), and more production took place in urban factories for goods like textiles, clothing, and machinery. A growing system of banks, insurance companies, and stock exchanges reflected the increasing financial complexity of the American economy. A new middle class of professionals and a large group of industrial laborers, including many immigrants, arose.

The growing capitalist economy was increasingly specialized and commercial. The self-sufficient yeoman farmer became a thing of the past. Instead, Americans were producing goods for larger markets, and workers were increasingly relegated to single, specialized tasks. While the economy was very productive, it was also susceptible to economic booms and busts. The Panic of 1837 was one example of the instability of the new economy.

Whenever possible, your answer should tell a story, following chronological order and linking events in cause-and-effect relationships.

Reform movements tried to improve the lot of people who lost their traditional sense of community. Evangelistic preachers like Charles Finney were extremely popular, especially in areas where there had been rapid economic change, like along the Erie Canal. These preachers stressed that people could create their own salvation. The religion was emotional, but also stressed the need to help reform society on a practical level by helping others. Other religious groups formed utopian communities that stressed helping each other and working together, unlike the competitiveness of capitalism.

Reform movements could also be used to deal with the problems of the growing working class. Many middle class reformers feared social disorder if workers became too unruly, and formed temperance societies to try to stop the widespread drinking. Other reformers, like Horace Mann and Dorothea Dix, tried to help solve the problems of the growing cities by reforming educational systems and insane asylums, respectively. Middle class reformers, like William Lloyd Garrison and Frederick Douglass, also fought for the abolition of slavery, while Elizabeth Cady Stanton and others fought for women's rights.

The first half of this essay question is fairly basic, although important for understanding the era. The important theme in this context is the sense of cultural dislocation economic change can bring. The second half is more difficult because it asks the student to connect two different aspects of history and show the connection between the two. A good answer displays organization, examples of the variety of reform movements, and clear connections between economic and cultural change.

2. At an 1830 presidential banquet in Washington D.C., in the midst of the nullification crisis, Andrew Jackson raised a toast, "Our Federal Union—It must be preserved." His vice president, John C. Calhoun, countered with the toast "The Union—next to our liberties most dear."

   What did these toasts mean, in the context of their era? What national tensions did these toasts reflect, especially considering the sectional crisis that led to the Civil War? Provide examples of this sectional tension and show how they illustrate these competing viewpoints.

I strongly advise students to leave themselves plenty of space for taking notes, and to approach each lecture understanding that there are two or three main ideas to each lecture. If you understand the main ideas, you will understand how the information presented in lecture is used to strengthen those ideas. To that end, I also advise students periodically to review their notes so they can avoid cramming and losing sight of "the big picture."

ANSWER

Jackson's and Calhoun's competing toasts reflected their views on the power of the federal union versus the states. Under Jackson, Congress enacted a protective tariff that benefitted Northern industry and hurt Southern agriculture. South Carolina, Calhoun's home state, rejected the tariff and asserted that states had the right to reject any federal law they found unconstitutional. Although South Carolina later compromised, these competing ideals of federal versus state power (which had existed from the birth of the nation) characterized the sectional struggle over westward expansion and slavery that led to the Civil War.

An excellent first paragraph that contains the essence of the entire answer. The phrase *competing toasts* is well chosen, thoroughly accurate, and demonstrates—instantly—the student's grasp of the significance of the quotations cited in the question.

The North and the South had different agendas and different ideas about what the federal government's role should be. The North, dominated by industry, favored a strong central government and policies that promoted industry such as tariffs and internal improvements. The South, dominated by agriculture and possessing a slave labor force, favored power in the hands of the states, as a way to protect slavery and economic stability. As the nation expanded westward and new states applied for statehood, these two competing economic systems and political philosophies clashed.

In 1820, when Missouri applied for statehood, Southern politicians rejected suggestions that Missouri be a free state. To protect the even balance of free and slave states, Maine was admitted as a free state at the same time Missouri was admitted as a slave state. In 1850, when California was admitted as a free state, Northern states had to make concessions such as the Fugitive Slave Law, which required Northerners to return runaway slaves to their owners in the South. In 1857, the Supreme Court's controversial Dred Scott decision ruled that slaves had no right to sue in courts of law and that Congress could not ban slavery in Western territories.

The preceding paragraph discusses political compromises between the North and South, setting up the discussion, in the next paragraph, of how the compromises and conflict developed and played out.

These sectional tensions were dramatic and even violent. The 1836 "Gag Rule" didn't allow discussion of abolition on the floor of Congress, although the antislavery movement was gaining momentum. Harriet Beecher Stowe's *Uncle Tom's Cabin* was an influential book that led many to sympathize with abolitionists and slaves. Preston Brooks, a senator from South Carolina, even savagely beat Northern Senator Charles Sumner when Sumner insulted another Southern senator.

Compromise between these two systems seemed less and less likely, especially after violence in Kansas and then again during John Brown's raid in Harpers Ferry, Virginia, in the 1850s. The North saw itself as a united nation based on bustling commerce and free labor, while the South relied on states' rights to protect not only its economic system, but its conception of a social ideal. When the two-party system broke down in the 1850s and Abraham Lincoln was elected as president on purely sectional lines, the Southern states felt their rights as states had been violated. They could not consent to a president who did not reflect their ideals or objectives in any way, and secession was the result.

This essay question asks the student to make a connection between an event from the Jacksonian Age and the ensuing development of the sectional crisis—a potential challenge. The student must look at the question and realize that the quotation and event represent a larger struggle over the role of government (a dominant theme of an early American survey course).

The description of the sectional crisis provides numerous examples of conflict, but, more importantly, they are organized into three paragraphs. The first paragraph (the third paragraph in the essay) discusses political compromises between the North and South. The second paragraph gives examples of how the conflict was played out. The final paragraph in the essay talks about the breakdown of political compromise. Importantly, the final few sentences tie this breakdown back to the original point of the essay, which was to discuss the relationship of the sectional crisis to these competing versions of how a government should rule.

3. Describe the variety of slave life, relationships with their masters, and the cultural institutions that slaves set up themselves. How did slave culture affect the sectional crisis and the Civil War? What were the typical goals of freedmen in Reconstruction, and were these goals achieved?

ANSWER

There was no one slave experience. The region, the nature of the work, the size of the master's home, and the master himself were all factors that affected the slaves' lives. Slave life could be brutal, but slaves nevertheless forged their own culture that helped them cope with their status and forge self-esteem and pride. Slave culture and a maintenance of a degree of autonomy helped blacks achieve limited gains during the Civil War and Reconstruction.

If the first paragraph is key to a successful essay response, an effective first sentence is crucial to a successful opening paragraph. This one begins by defining slavery as a multi-faceted experience—a concept that was emphasized in course lectures, so that the very first sentence demonstrates the student's command of an idea central to the course.

Most field slaves worked long hours, especially in the summer. On large plantations, slaves often worked in the gang system, while skilled workers often received opportunities to work independently on tasks. Others worked in house jobs like maids, cooks, and drivers; these jobs had better conditions, but these slaves were also more removed from the larger community of field slaves. Clothing, conditions, and food were usually less than desirable.

Use specific detail wherever possible. Avoid vague abstraction. The preceding paragraph is vivid.

Slave codes and discipline were tighter by 1831, after Nat Turner's unsuccessful revolt and the rise of abolitionist pressure in the North, such as from William Lloyd Garrison. Slaves could be rigidly supervised, and state governments set conditions against freeing slaves and allowing blacks to congregate without white supervision. While most masters encouraged slaves to form families because it would discourage runaways, slave families could also be broken up for economic reasons. While many slaveholders were probably not brutal and inhumane, the institution of slavery was undoubtedly evil and could have terrible consequences for blacks.

This material about slavery after 1831 is nicely presented; however, there is no specific indication here as to what slavery was like before 1831. This does not fatally flaw an essay written under the pressure of time, but it does serve to remind us that if we talk about what happened *after* a certain milestone or date, we need to compare it to what happened *before* as well.

Slaves tended to balance the problems inherent in slavery by creating their own culture that encouraged autonomy and subtle resistance to masters. Through families, peer groups, songs, and stories, slaves taught each other lessons about black solidarity, ways to deal with white power, and how to maintain self-esteem and a desire for freedom. They also engaged in various forms of resistance such as acting sick, breaking tools, or burning crops. While the power of the master was strong, it was not absolute.

Slaves' maintenance of pride and a desire to forge their own destiny influenced the course of the Civil War and Reconstruction. By the Civil War, as Union troops infiltrated the South, slaves flocked in large numbers to Union lines. By making themselves an issue for the Union to deal with, they influenced Abraham Lincoln's decision to issue the Emancipation Proclamation in the fall of 1862. Blacks also expressed strong desires to participate in the military effort, which many did with great success. Toward the end of the war, blacks resisted work contracts that closely approximated slavery, and actively sought sharecropping and wage contracts that gave them more autonomy than gang labor tasks.

Freedmen achieved limited gains during Reconstruction. While there were temporary increases in political participation, black rights were not protected and their political voices were silenced by brutal repression. In the long term, the federal government failed to protect blacks' right to vote, their physical safety, and their right to a non-discriminatory workplace, with tragic consequences.

This essay question asks the student to draw out the broader theme of black agency throughout the nineteenth century, but also to realize its limits due to the systems blacks operated under. It requires a less traditional perspective on fairly well-known institutions and events.

The most important concepts are the variety of slave experiences, the existence and ramifications of slave culture, and blacks' role in the Civil War and Reconstruction. A well-done essay brings these concepts together through a strong introduction and body paragraphs that address these issues separately, while weaving through the common theme of the extent of black agency.

4. What did the Confederacy hope to achieve by seceding? What were the war aims of the Union? How did each side's war objectives influence their military tactics? Why did both sides believe they would win? Why did the Union eventually win?

ANSWER

Entering the war, both the North and the South believed that they were preserving national ideals of freedom and liberty. Both sides also believed they would win; the South would win just by holding off a Northern invasion, while the North believed it had superior resources and manpower. Only by the extension of the war into a long, brutal conflict could the North eventually win.

An excellent encapsulating first paragraph. Don't begin an essay with vague background remarks. Instead, dig right into the question. If possible, *answer* the question (or at least the main part of the question) in the very first paragraph, then use the rest of the essay to elaborate on and to support your answer.

The Southern states did not necessarily need to secede from the Union when they did. Abraham Lincoln had publicly stated that he would not try to destroy slavery where it already existed. However, Lincoln had steadfastly maintained that slavery should not be extended into the new territories. When Lincoln was voted in as president despite a total lack of support in the South, the Southern states felt that their long-term interests in protecting the social and political institution of slavery were threatened. Especially in the context of rising abolitionist pressure and sectionalist violence (like John Brown's raid on Harpers

Ferry), the South believed that through secession, they could preserve slavery, in effect preserving their way of life and protecting the liberty of white men.

Use paragraph structure to break thoughts into logical blocks. The preceding paragraph is devoted to the South. The next paragraph, to the North.

The North did not originally fight to destroy slavery, but rather to save the Union. They could not just let the Southern states secede if they wanted to protect the stability and legitimacy of the national government. Since a stable national union was necessary to protect the American constitutional republic, the North thus felt they were also fighting to preserve the liberty of white men.

Although each side believed it would win quickly, a variety of factors prolonged the war. Northern generals like George McClellan repeatedly failed to aggressively attack Confederate armies when they were weak. While the Confederacy was mostly successful in defending its territory in Virginia, it failed in its two major forays into Northern territory, Antietam and Gettysburg. At the same time, Union armies were inching southward in the Western theater of the war.

The weakest features of most essays are poor transitions from one paragraph—and thought or topic—to the next. An essay is a *linked* narrative, not a laundry list of separate ideas. Transitions are the links between paragraphs and ideas. Take a close look at the opening sentence of the preceding paragraph and note how it moves from the two paragraphs that come before it (each devoted to a different side) to a new topic: the factors that prolonged the war.

The extension of the war hurt the South in particular, since the industrial and financial resources necessary for war were located in the North, and Union ships had blockaded Southern ports. While each side had significant internal tensions (for instance, Lincoln was barely reelected in 1864), both sides' official resolve did not weaken. A superior infrastructure, shrewd political leadership, and improved military leadership helped the Union win. At the same time, the Union broadened its definition of liberty by issuing the Emancipation Proclamation. This act resolved the question of how to handle the increased pressure from slaves on the Union army, helped open the door for black military participation, and gave the Union troops a new moral legitimacy.

This question hinges on a few major ideas, especially the various conceptions of liberty and freedom. Even though the question does not deal specifically with the preceding sectional

crisis, it does ask the student to illustrate how the South could justify secession, and the North military action.

The second major idea is how the North actually won the war. It is important to realize that Union victory was not a foregone conclusion. More basically, one should not argue that the war was originally designed to determine the fate of slavery in the South. The student should grasp how emancipation became a war aim only due to the changing circumstances of the war.

Students are often fascinated by the Civil War and some tend to know more about it than any other period in an early American history survey course. There are numerous details students may bring in to illustrate these concepts, but they should remember above all else what motivated each side, and how these motivations were altered by the exigencies of the war.

CHAPTER 24

UNIVERSITY OF SOUTH CAROLINA

HIST 111: U.S. HISTORY TO 1877

Kevin Gannon, Graduate Instructional Assistant

THE GOAL OF THIS COURSE IS TO IMPART A BASIC UNDERSTANDING OF EVENTS, ISSUES, themes, and actors significant to U.S. history through Reconstruction, and to develop skills that will be necessary in the college careers of my students, such as critical analysis and clear writing. A central skill is the ability to argue a point well in a clearly written essay, articulating ideas in an intelligent and coherent manner, and analyzing sources to draw informed and critical conclusions.

The midterm for this course covers roughly the colonial period through the 1820s.

MIDTERM EXAM

Section 1: Chronology. *Arrange the following events in the order in which they occurred:*

1. Thomas Jefferson is elected president.
2. The Puritans found the Massachusetts Bay Colony.
3. The Erie Canal is completed.
4. The US goes to war with Britain for a *second* time.
5. The Constitution is ratified by enough states to put it into effect and bring the U.S. a new government.

6. The Stamp Act's passage triggers significant colonial protest.
7. The Alien and Sedition Acts are passed.
8. The Seven Years' War (or French and Indian War) ends.
9. A burst of revivals known as the "Second Great Awakening" occurs.
10. Eli Whitney invents the cotton gin.

Answers (earliest to latest): *2, 8, 6, 5, 10, 7, 1, 4, 3, 9*

Memorizing exact dates of events is far less important to an understanding of history than developing a grasp of approximate date and sequence.

Section 2: Matching. *Match the people in List A with the description or item in List B most relevant to them.*

List A

1. John Quincy Adams
2. John Winthrop
3. Robert Fulton
4. Thomas Hutchinson
5. Francis Marion
6. Denmark Vesey
7. James Madison
8. Andrew Jackson
9. Charles Grandison Finney
10. John Smith

List B

a. leader of guerilla forces in South Carolina during the Revolution
b. first governor of Virginia colony
c. had perhaps the most influence on the form of government embodied by the Constitution
d. Secretary of State for Monroe, eventually president in his own right
e. popular hero of the War of 1812 from his leadership at the Battle of New Orleans
f. governor of Massachusetts during the years of crisis leading to the Revolution
g. leader of a large, though unsuccessful, conspiracy aimed at a slave rebellion

in South Carolina

h. most famous revivalist preacher during the Second Great Awakening

i. Puritan governor of the Massachusetts Bay colony for its first twenty-five years

j. inventor of the steamboat

Answers: *1-d, 2-i, 3-j, 4-f, 5-a, 6-g, 7-c, 8-e, 9-h, 10-b*

My exams are heavily weighted toward the essay portion. The objective portion of the test (chronology and matching sections) is designed to ensure that students have the basics down, which is necessary in order to present a well-argued essay in the remainder of the test.

Essay Section:

In the essay portions of my tests, I try to pick a large, central issue from the section of the course covered by the exam to structure an essay around. In this case, that issue is the American Revolution as the result of a "perception gap" between the colonies and Britain. I tend to use this theme as one way of looking at the Revolution in my class, and it is a useful one, because it prompts students to consider not only the immediate cause of hostilities (shots fired in April 1775), but the long-term breakdown in colonial-British relations as well. In their responses, students need to address, first, whether they believe there was a "perception gap" and why, then, second, they must provide evidence to justify their answer. A good answer considers events as far back as the close of the Seven Years' War, and how issues left unresolved affected the course of events throughout the next couple of decades. A successful response would also address the motives behind British colonial legislation (most importantly, mercantilism), and the reasons many colonists reacted to the legislation in such a hostile manner. Specific dates or a "laundry list" account of British taxation policy is not the main objective of the question. The steady deterioration of relations and the growth of this perception gap are what I really want to see explored here.

Some historians argue that the American Revolution is best seen as the result of a "perception gap" that grew between the colonies and the British government, which eventually culminated in armed conflict. In a well-argued essay, assess the validity of this idea, and provide supporting evidence for your claims.

ANSWER

The American Revolution was indeed the result of a "perception gap" between the colonies and Britain. The government in England had its reasons for doing what it did, but they didn't understand how the colonists were going to react. By the time the Revolution broke out, this gap had grown so wide that the only way for the issues to be resolved, according to the colonists, was by separating from Britain.

The question instructs you to begin by answering whether or not a "perception gap" existed. Make certain that you answer this question directly and unequivocally. Then go on to define the perception gap.

To truly understand the origins of the "perception gap," one must start at the end of the French and Indian War. In 1763, the peace treaty that ended the war gave Britain almost complete domination in North America, because the French lost almost all of their colonies. However, Britain also had a *huge* debt coming out of the war. Since a lot of the action took place in North America, to protect the British colonists there, the British government decided that those colonists should help pay off the debt. To do this, the government decided to begin taxing a number of colonial goods to provide revenue and pay off the war debt.

The structure of this answer is logical. After answering that a perception gap existed, the student defines that gap, then proceeds to an explanation of the origin of the gap. Allow the question to focus your response; in this case, don't let go of the perception gap concept.

The first of these taxes was the Sugar Act, which came shortly after the war ended. This was important because it taxed an important part of colonial trade, molasses (made from sugar), and was bound to provoke criticism. The act also enlarged the power of the courts of trade to administer the law and punish violations, something that was not popular with a lot of colonists. In Boston, for example, there were protests over this act.

While the Sugar Act was controversial, the Stamp Act that came in the next year was even more so. The British government decided to tax official documents, playing cards, newspapers, etc. by requiring that a wide variety of these documents bear an official stamp, which had to be bought from a special agent. This really set off the colonists. There were riots, stamp commissioners were tarred and feathered, and protests erupted all over the colonies. Most significantly, however, the colonies organized to try to present a united front to protest the act. A "Stamp Act Congress" was called, and representatives from different colonies met to draft a petition to the King to get him to repeal the law. Also, an organized boycott of British goods took place. Faced with this unanticipated response, the British government backed down and repealed the act about a year later. But they also passed the Declaratory Act, which basically said that Parliament still had ultimate authority to legislate for the colonies.

Be specific wherever possible. Tell a story. Review real—and relevant—events. This is all *evidence.* Avoid unnecessary abstractions or vague opinions. These are not evidence, but *hot air.*

This was a significant step in the development of a perception gap. Britain felt that since the protection of the colonists was the ultimate goal of the recent war, that the colonists should realize what a sacrifice the mother country made and help to repay the debt. However, the colonists saw things differently. Many of them also fought beside the British, and the colonists felt that they had sacrificed as well; therefore, they should not have to bear the entire burden, as it seemed Britain was forcing them to do. The colonists' main grievance, though, was how they weren't represented in Parliament, the body that passed laws. Since they had no voice in these matters, the colonists felt like they were being deprived of a fundamental right that other British subjects enjoyed. "No taxation without representation" became the rallying cry. Britain's answer to this was the idea of "virtual representation," which said that Parliament represented ALL British, whether they voted or not—that they were acting in the best interests of the entire realm. So, in theory, even though the colonists had no vote for representatives, they were still represented because they were British subjects. The fact that the government largely expected the colonists to buy this argument is evidence of how far the gap had grown already by this point.

While the student uses actual events as valuable evidence, she doesn't let the response drown under them. The preceding paragraph underscores the cause-and-effect significance of the events. History often focuses on issues of cause and effect; answers that identify cause-and-effect relationships are typically quite successful.

Other taxes and measures would follow, such as the hated Townshend duties (including the Tea Act, prompting the famous Boston Tea Party), and the colonial protests became even stronger. Again, the themes were largely the same: virtual representation versus no taxation without representation. Neither side could be convinced that the other's point of view was correct. Further fuel to the fire was added by the Quartering Act, which dictated that colonists had to support British troops on their soil. This was controversial because a standing army was seen as a big threat to liberty by the colonists. The situation was worsened by the "Boston Massacre" of 1770, where British troops (probably in self-defense) fired into a mob of Bostonians one winter night. While the motive may not have been murder, the propaganda generated by the situation inflamed anti-British sentiment even further. The Intolerable Acts were another example of British policy backfiring. The acts, which were passed in retaliation for the Boston Tea Party, actually aroused more anti-British feeling than ever before. Even though the British government was trying to punish Massachusetts, and make it an example, what actually happened is that another step toward colonial unity was taken when the colonies helped support the people of Boston while the port was closed.

That same year, 1774, was when the First Continental Congress was called together. This group met to organize a nonimportation agreement against Britain, because many colonists felt that the British government was ignoring their demands for representation while still passing unfair laws. The British really didn't realize how separate the colonists already thought themselves—many referred to themselves as "Americans" rather than "British." By the time shots were fired at Lexington and Concord in April of 1775, the Revolution was almost a given. Because of this process by which Britain ignored or failed to realize the seriousness of American demands, the two parties kept drifting further apart. It really began to seem to many colonists like their was some sort of British conspiracy to deprive them of their rights. Because the British failed to understand this colonial perception, they unwittingly aided this sentiment in the colonies by the laws Parliament passed. The growth of this "perception gap" is one of the fundamental reasons that the Revolution took the form it did, when it did.

This successful essay takes pains to point out not just that an event was significant, but *why* it was. This is a crucial part of the answer's success.

Note how the entire essay allows chronology of events to dictate the form of the response. Chronology is usually the most natural way to structure a history essay response. Also note how the student is careful to return to the perception gap theme in the concluding paragraph. Drive your point home. Leave no loose ends. Do not leave it to your reader to draw the conclusions you should draw.

FINAL EXAM

In my classes, the final exam is not usually cumulative; that is, it goes back only to about the Constitutional period rather than the colonial beginnings of the nation. While I believe many professors use this approach, probably even more make their final exams cumulative. The choice depends in large measure on the amount of time allotted for the exam. I prefer to have students spend a lot of time on the big interpretive essay portion of this exam, which deals with slavery as a major issue in the coming of the Civil War. This question forces them to consider evidence stretching all the way back to the Constitution's ratification, and it takes some time to answer.

Section 1: Chronology. *Put the following events in their proper order, earliest to latest:*

1. Abraham Lincoln is elected president.
2. Andrew Jackson "wins" the "war" against the Second Bank of the United States.
3. Nat Turner leads the bloodiest slave rebellion in U.S. history in southern Virginia.
4. A new, tougher Fugitive Slave Act is passed.
5. The Mexican-American War begins.
6. South Carolina Confederates open fire on Fort Sumter.
7. Gold is discovered in California.
8. The Kansas-Nebraska Act is passed.
9. Lincoln and Douglas carry out their famous debates in Illinois.
10. John Brown leads a raid on the federal arsenal at Harpers Ferry, Virginia (now West Virginia).

Answers (from earliest to latest): *3, 2, 5, 7, 4, 8, 9, 10, 1, 6*

While specific dates aren't as important as analysis, knowing when events happened in relation to each other is of crucial importance for that analysis; hence the chronology section of this exam. For example, students should understand that the discovery of gold in California led to a crisis over its admission into the union, which precipitated the Compromise of 1850, which included the newer, tougher Fugitive Slave Act. I am a big believer in understanding the importance of cause-and-effect relationships.

Section 2: *Fill in the blank with the appropriate answer:*

1. Abraham Lincoln and Stephen Douglas held their famous debates in a Senate race in Illinois. The winner of that election was ______________. **Answer:** *Douglas*

2. The theory that South Carolina used to assert its right to ignore federal tariff laws, a theory that was articulated by John C. Calhoun, was called ________. **Answer:** *nullification*

3. The phrase ______________ was used to describe the widespread belief during this period that America was ordained to spread across the entire continent and that the American way of life was somehow superior. **Answer:** *Manifest Destiny*

4. In 1857, Chief Justice Taney wrote the Supreme Court opinion for this highly controversial case involving a slave suing for his freedom:______________. **Answer:** *The Dred Scott Case*

5. Harriet Beecher Stowe's novel, ______________, touched off a firestorm of controversy over the slavery issue in the 1850s. **Answer:** *Uncle Tom's Cabin*

6. This fiery abolitionist was the editor of the *Liberator* and among the most radical of the antislavery activists: ____________________________. **Answer:** *William Lloyd Garrison*

7. The driving force behind both the Missouri Compromise of 1820 and the Compromise of 1850 was ______________. **Answer:** *Henry Clay*

8. ______________ was the first state to secede from the Union following Lincoln's election. **Answer:** *South Carolina*

Don't get caught up in memorizing every name and date in the course. While names and dates are important, it's even more important to be able to prioritize information—that is, to be able to realize the overarching themes and issues that become significant forces in a given historical period. In short, students can utilize a number of reference sources to find out, say, who ran

against Andrew Jackson in 1828, but it is much more difficult, yet more valuable as well, to be able to understand, for example, why slavery became such a crucial political and social issue in the second quarter of the nineteenth century. Most instructors would rather their students be able to provide this sort of analysis rather than regurgitate rote-style timelines and the like.

Essay Section

Historians have debated for years the idea that slavery caused the Civil War. Having just completed History 111, and being now qualified to weigh in with your judgment, consider that idea. In a well-constructed and vigorously argued essay, give your assessment of the idea that slavery was the cause of secession and war.

This essay topic is meant to be a synthesis of the second half of the course. By assessing the impact of slavery on the coming of the Civil War, students really have to consider much of what was covered in the course's second half. The successful answer generally argues that slavery, while not the *only* cause of the war, was certainly a major (or the major) cause of secession and hostilities. It will also prove this assertion by going back at least as far as the Missouri Compromise, where slavery once again becomes a national issue. The truly excellent essay will discuss the controversy over slavery during the period of ratification of the Constitution. An essay that goes back only to 1850 (or even worse, the election of 1860) does not begin to answer the question successfully. One of the most significant themes of antebellum U.S. history is the growth of the slavery issue into a matter that becomes almost totally irreconcilable by 1860. This growth took time, and there were many steps, compromises, and other attempts to stem the discontent along the way. This essay is meant to prompt students to reflect upon that.

ANSWER

Slavery was a major theme in the nineteenth century for Americans. Even though the institution did not really exist in the North any more, the whole country—North and South—was tied up in it. Slavery was the mainstay of the Southern economy, and Northern merchants and manufacturers profited indirectly from it as well, as Southern cotton became a big part of the national economy. Therefore, arguments to end slavery would meet resistance from northerners and southerners alike, from both economic and prejudice-related reasons. Yet, over time, opposition to slavery became greater and greater. This caused those in favor of slavery to dig in their heels even more. The end result of this intractability was the Civil War. Slavery was the one question that remained unanswered during the antebellum period. While there were other reasons southerners seceded—states' rights, property rights, a feeling that their liberties were being violated—many of these really rested on the foundation built by slave society. The issue was compromised, ignored, etc., but it never really went away. Eventually, the question would be answered by four years of war.

The student might have declared in his first sentence that slavery was the chief cause of the Civil War. This would have been acceptable. He chose, however, to pave the way with a brief introduction to the significance of slavery. The important point is, however, that he *did* answer the specific question asked, and he did so in the first paragraph, if not in the very first sentence. Often, providing background evidence is a good way to begin a response. Just make certain that the background doesn't bury your response to the core of the question!

Even when the new government of the U.S. was set up by the Constitution, slavery was an issue. Southerners were afraid that the new document would threaten their slaves' status, and many northerners were unhappy about the idea of the new government protecting slavery in a supposed "land of the free." Eventually, the first of many compromises was created, the Three-fifths Compromise, which stated that slaves would count as 3/5 of a person for purposes of taxation and representation. This ensured that southerners would have congressional representation equal to the northerners, and that one section would not dominate the other. This became especially important after 1793. In that year, Eli Whitney invented the cotton gin, making the growing of cotton significantly easier and much more profitable. The result was the spread of cotton agriculture, which was dependent upon slave labor, all across the South. The hope of the founding generation that slavery would gradually fade away within a few decades would never be realized, as cotton propelled the South into one of the world's richest regions, and became a major staple in the entire nation's economy.

For a while, slavery was, if not accepted by all, at least left undebated. When new states were admitted to the Union, they came in pairs—one slave and one free. This kept the balance in Congress safe. However, in 1819, Missouri applied for admission as a slave state, and a crisis erupted as it looked like the balance would be upset. Some northerners had been arguing since the Louisiana Purchase that the new western lands would be slave-filled, and the North would be drowned out by slave states, and it looked to many like they were right in 1819. Eventually, a compromise was brokered by Henry Clay of Kentucky, but not before the issue of slavery came right back onto center stage. Thomas Jefferson said it reminded him of a "firebell in the night"—that the alarm over slavery had been rung, and it would be hard to extinguish the coming blaze.

Quotation from contemporaries is often a very effective element of a thoughtful answer. Note that the student does not quote Jefferson only to show that she can do so. The next paragraph develops Jefferson's metaphor further. Thus the student shows not only that she can quote Jefferson, but that she understands the significance and the concepts that underlie the quotation. This demonstrates a genuine engagement with the material at hand—something that is always warmly welcomed by instructors.

This became a more appropriate metaphor with the rise of "immediatist" abolitionism in the early 1830s. The failure of gradual solutions, like colonization, was frustrating to the small core of antislavery activists in the U.S. When Great Britain ended slavery in its empire in the 1820s, American abolitionists began to take heart. The immediate symbol of the transition from "gradualism" to "immediatism" was the appearance of William Lloyd Garrison's *Liberator*, where he announced that the time for "wishy-washy" sorts of solutions was over, and that the national sin of slavery needed to be eradicated.

Southerners met this new sort of abolitionism with alarm. They acted to table antislavery petitions in Congress before they could be read (the "gag rule"), and agitated for abolitionist mailings to be suppressed in the South, which was eventually ordered by President Jackson. The room for compromise became smaller and smaller as the stakes became higher. The South's ever-increasing prosperity was tied directly to slavery, and many northerners had a direct economic stake in its continuance as well. Yet there were also many, like the abolitionists, who pointed out the contradiction of the existence of slavery in a land dedicated to freedom and democracy—a contradiction that seemed to be becoming harder to ignore.

The essay proceeds chronologically—the simplest, most direct way to show the development of a situation, crisis, or idea. Just be aware that structuring a response this way requires knowledge of the chronology of events. If this knowledge is faulty, the essay will be badly flawed.

The question became even more urgent with the acquisition of the territory from the Mexican War. The question of whether the territory would be slave or free troubled Americans even before the territory was officially added to the country. The Wilmot Proviso of 1846 was introduced into Congress, which stated that all of this land would be banned from having slaves, which started an immediate controversy. The proviso didn't pass, but the controversy surrounding it remained, as southerners and northerners began to distrust the motives of the other side more and more. What to do with the territory became extremely important in 1849, when gold was discovered in California. The rapid increase in the population made California ready for statehood by 1850, which set off yet another crisis, as California became a rich prize coveted by both North and South. Eventually, another compromise was put together by Henry Clay, with the assistance of Stephen Douglas. This compromise was meant to appease both sides, with California entering as a free state, a new and harsh fugitive slave act was passed, and the rest of the Mexican cession left undecided. Even though it was supposed to make everyone happy, it ended up agitating the issue even more. Many southerners felt that too much was given up (like California) in the compromise, while the Fugitive Slave Act triggered a storm of protest in the North. By requiring northerners to help capture runaway slaves, and depriving the accused of a jury trial, the act turned many northerners who were previously ambivalent about slavery into antislavery supporters. It brought the issue home to many who had previously seen themselves as having no direct stake in the conflict, which would be very significant in the coming decade.

An even more drastic step toward war was taken in 1854 with the passage of the Kansas-Nebraska Act. The act abolished the Missouri Compromise line, which was still thought by many to be the main route of compromise between slave and free states, and enacted popular sovereignty in its place. Popular sovereignty was the idea that the people of a territory got to choose whether they would be a slave or free state. This sounded nice, but it was too vague: When would they choose? How many people were necessary to hold a referendum? Unanswered questions like this led to a virtual civil war in Kansas as both pro- and anti-slavery forces sought to make that state either slave or free. The Kansas-Nebraska Act also basically destroyed the two-party system of politics, because the Whig party split over its passage (southern Whigs were for it, northern against it). By removing this party system, the way was clear for formerly isolated third parties to gain national strength. The main one was the Republicans—the first major U.S. party to take a firm antislavery stance, and the only major party to be purely sectional in nature. The existence of this party heightened already serious southern fears of political domination by the North—ironic, considering that many in the North feared the same thing from the South—a sentiment made stronger by the Dred Scott decision of 1857, which seemed to remove any barriers against slavery's expansion. Roger Taney declared that Congress had no right to interfere with slavery, and many northerners feared that slavery would now expand unchecked across the country.

The final straw for the South was the election of Lincoln to the presidency in 1860. A purely sectional party electing a purely sectional candidate was too much, southerners argued—especially if that candidate was an antislavery man. Regardless of the fact that the election was constitutionally proper, the southern states began to secede in the winter of 1860-61, led by South Carolina. The arguments for secession were various, such as sectional political power had overwhelmed the South, or that their property rights were threatened, or their liberties and state rights were being trampled. Yet these arguments all related to the issue of slavery, and thus while they were reasons for war, they were subsidiary to the overarching issue of slavery. When southerners agitated about property rights, they were talking about slave property. The issue of state rights often connected with the issue of federal interference in the institution of slavery. And a main reason for the sectionalization of politics was over slavery, particularly the debate over slavery in the new western territories. Therefore, while slavery was not the only cause of Civil War, it was the root of the controversy. When the South seceded after the election of 1860, it was largely the culmination of the decades-long growth of the slavery controversy.

The essay concludes by repeating the main point—that the Civil War was indeed caused by slavery. Note that the student does not *ignore* other causes, but does show that they were subsidiary to the "overarching" cause of slavery. A good response does not overlook opposing points of view, but takes them into account, showing how they relate to the point of view advocated in the response.

CHAPTER 25

VALPARAISO UNIVERSITY

U.S. HISTORY 92D: CIVIL WAR TO PRESENT

Stephanie E. Yuhl, Post-Doctoral Fellow

A MAJOR OBJECTIVE OF THE COURSE IS TO TEACH STUDENTS TO DEVELOP CRITICAL thinking and effective communication skills while deploying historical data from the American experience. Students are also encouraged to be broad-minded and sensitive in their approach to the variety of ways in which people experienced the past. They are also encouraged to respect each other's ideas and claims, while nonetheless feeling free to challenge each other, as long as they have data to back up their opposing claims.

A key skill this examination demands is effective communication. Another skill is the ability to discern for onself what information is pertinent to the issue at hand. I discourage students from merely regurgitating every last bit of info they can remember about a given topic.

MIDTERM EXAM

Short Essay *(40 points):*

The Fifteenth Amendment to the Constitution created problems for the women's rights movement. In a brief but informative essay, explain the causes and results of this crisis. Was it avoidable?

ANSWER

As part of the post-bellum Reconstruction of the Union, Congress passed several amendments to the Constitution aimed at aiding the transition of former slaves to their freedom. These included the 13th (emancipation), 14th (citizenship), and 15th (right to vote). Many participants in the women's suffrage movement had long been advocates of freedom for blacks, and some activists such as Elizabeth Cady Stanton and Lucy Stone, had even cut their teeth in the public sphere as Abolitionists. Congressional extension of the vote to newly emancipated black men only precipitated a crisis in the ranks of the women's movement, however. The result was a split women's movement, which was likely unavoidable given the strong personalities of those involved and the fact that America at the time, despite the emancipation of the slaves, was still a place where blackness was considered inferior to whiteness. Furthermore, this split illustrates the difficulties of creating a social movement that embraces people from a variety of conditions and perspectives.

The split in the women's movement is of complex origin. This opening paragraph does a very good job of laying out the background of the movement, then tracing the origin of the split. The concluding sentence places the *particular* problems of the women's movement in the context of the problems social movements in *general* face.

In 1869, some white women activists resented the government's privileging the franchise concerns of black men over those of women. The leaders of this faction included such longtime activists as Elizabeth Cady Stanton and Susan B. Anthony, who declared such an amendment would create an "aristocracy of sex" in this country. They saw in the 15th Amendment debates a chance to reassert female suffrage to the national agenda, which had long taken a back seat to issues concerning African Americans during the Civil War.

Other women's rights activists, such as Frederick Douglass and Lucy Stone, saw the 15th amendment as a necessary step in the continued abolition of all minority oppression in the U.S., which would lead eventually to female suffrage. It was, they argued, "the Negro's Hour," and the Amendment should pass as written with the words "male suffrage." The time for women would follow. This split in viewpoints between immediate versus gradual change is a common theme in the history of American social movements.

The opposing viewpoints within the women's movement are clearly explained—one brief defining paragraph devoted to each. In exam essays, it is often most effective to define a single point of view in one paragraph and devote a new paragraph to an opposing point of view rather

than discuss opposing points of view in a single long paragraph. Thoughtful paragraphing underscores the direction of your argument. Do whatever you can to make your discussion clear.

As a result of this difference of opinion, the women's suffrage movement split violently into two groups with the same end in sight, but with very different strategies for achieving it: the American Women's Suffrage Association (AWSA) and the National Women's Suffrage Association (NWSA).

Again, a brief paragraph is used to highlight a specific instance of cause and effect. If this were a published essay, such brief paragraphing might be interpreted as clumsy or "choppy." In a time-constrained essay, however, breaking your work into blocks of thought is an aid to *your* composition and to the *reader's* comprehension.

Stone's AWSA sought to win female suffrage through conventional means at the state level, while Anthony and Stanton's more radical NWSA fought at the national level through more flamboyant, non-Victorian tactics. For example, the NWSA supported free-love advocate Victoria Woodhull for president in 1872.

This divisive event reveals the problems inherent in an atmosphere when multiple social movements exist. During the late 19th century, the women's movement struggled with the labor movement (Populists) and the black civil rights movement for energy and attention. While this environment can be very stimulating, it can also breed contempt. Competing agendas sometimes bring to the fore nasty elements in progressive movements, such as, in this case, racism.

Supporters of the NWSA often made derisive comments about the black man's capabilities in voting as compared to the abilities of white women. They played this "race card" to sway white men to their way of thinking. The tensions between competing ways of identifying onself, as a woman or as a black person or white person, can sometimes create problems within movements that tend to be single-issue focused. This problem of who speaks for whom was evident again later during the feminist movement of the 1960s/70s and persists today.

Middle-class straight white women like Betty Friedan claimed to speak for all women, when, in fact, their experiences and needs differed greatly from women of different races and classes and sexual orientations. The NWSA/AWSA split was unavoidable, given its context and the fervor of its leaders. A century later, feminist scholar Bell Hooks pointed out the persistence of this problem of competing agendas with women's acitivism when she declared that there is no one feminist movement.

This essay effectively *describes* the split in the women's movement, but is quite weak on argumentation. The author points to themes common in social movement history, without elaborating on them in a forceful manner. Even though this is only a short essay question, a better-outlined essay is expected, including a thesis statement and a conclusion of more weight.

Identifications *(60 points—15 points each). You must answer all four identifications, being sure to explain their significance to U.S. history.*

1. The Confederate Tradition/Lost Cause

Answer: After the South's defeat in the Civil War, Confederate veterans organized social/fraternal groups bent on recalling and memorializing their experiences in the battlefield in an attempt to negotiate the difficult transition back into the victorious Union. The most vocal of these groups, the Confederate Veterans, held many public rituals, such as reunions, parades, picnics, statue unveilings, and even some reenactments. The Confederate Veterans also wrote up the history of the Confederacy to ensure that their views of the past were not overshadowed by those of their conquerors. This organization spawned others, whose membership comprised the sons and daughters of the Veterans, which continued the Veterans' early mission. Although originally a highly sectional tradition, which felt its way of life marginalized and jeopardized by the Union, in time the Confederate Tradition moved comfortably with similar Yankee traditions of the Grand Army of the Republic. Beginning at the end of the 19th century, vets from both sides held joint reunions, easing the way for sectional reconciliation. The Confederate Tradition, as a states rights/libertarian tradition, is experiencing a renaissance in reenactments, Confederate flag displays, and debates, and in the rhetoric of the Southern Right.

This is an excellent identification, bringing the themes of the Lost Cause into high relief in the present American cultural political environment. The student has moved well beyond the parameters of a simple identification, and that is applauded.

2. The Atlanta Compromise

Answer: This was the label applied to a speech made by the African-American leader Booker T. Washington in Atlanta in 1895. He advocated industrial arts education for blacks (instead of classical/intellectual education, which other leaders like DuBois advocated) as practiced at the Tuskegee and Hampton Institutes. He argued that once blacks proved their real economic worth to whites, racism would begin to disappear. Washington further angered black radicals and pleased accommodationist whites by supporting racial segregation in social life. He envisioned the segregated social relationship as a productive hand whose fingers were separate but nonetheless connected for the common good. This ideology was reinforced the following year by the Supreme Court decision in favor of the doctrine of separate but equal public facilities in *Plessy v. Ferguson*, which formalized the de facto Jim Crow South and led to the further deterioration of black rights, including suffrage. It also led to the formation of a more radical black voice in America a decade later in DuBois and the Niagara Movement. This group was opposed to Washington's gradualism.

All the necessary data are in this identification. More commentary on gradualism versus immediacy would have fleshed out the response better.

3. The Palmer Raids

Answer: After the First World War and the Russian Revolution, America experienced a wave of anti-Communist sentiment, which manifested itself most violently in the Palmer Raids of 1919. Named after the U.S. attorney general, the raids included rounding up anarchists, immigrants, intellectuals, and individuals thought to be suspicious, and jailing them. The civil rights of many Americans were violated in the process. The raids are significant because they reveal the extension of wartime fears of sedition and difference in this country, which extended to include the rise of the second KKK, immigration quotas, and eventually the Scopes Monkey Trial (1925). These raids ushered in a period in the twenties of conformity, assimilation, political conservatism, and challenges to labor and free speech.

For an even better identification, the student could have explored the relationship between intolerance and wartime fears or the disparity between social myth and social reality in the 1920s.

4. Reconstruction Finance Corporation

Answer: This was President Herbert Hoover's response to the Depression. He made money available through the government to resuscitate banks, railroads, and financial institutions to keep them from going under and taking the rest of the economy with them. By funneling millions in from the top and hoping it would trickle down, Hoover sought to stem decline. The scheme did not work, and instead Hoover was vilified by the working poor, who, dispossessed, were forced to live in shanty towns they called Hoovervilles. The failure of the RFC paved the way for the beginnings of government intervention into the welfare of its citizens with FDR's New Deal and signaled the end of Republican rule that had dominated the 1920s.

This identification covers all the bases and is therefore quite adequate.

FINAL EXAM

Answer four of the following five identifications for a total of 40 points (10 points each). Be sure to note the significance of the identification in your answer.

1. Sacco & Vanzetti

Answer: Sacco & Vanzetti were two Italian anarchists who were accused of robbing and killing two guards at a factory in S. Braintree, MA in 1920. They were found guilty in 1921 and sentenced to the electric chair. A large segment of society rallied behind the men, protesting that they received an unfair trial, that in fact they were being targeted for being immigrants and for their political beliefs more than for any act they might have committed. They died in 1927. The divisions of this case show the trends in 1920s America toward conservatism (Scopes and Red Scare) and xenophobia (immigration quotas) and intolerance (KKK).

To make this a better identification, the student should expand on his/her understanding of the divisions in America in the 1920s and what they indicate about the political process and about the nature of what it means to be an American. The response needs to situate this trial more explicitly in the context of the culture wars of the day, not just list some of its manifestations.

2. *Watkins v. the United States*

Answer: This case involved the appeal of labor leader John Watkins's contempt of court charge before the House Un-American Activities Committee. Watkins refused to "name names" of people he knew to be Communists and challenged the committee's right to force him to do so. His appeal to the Supreme Court was successful. The Court concluded that Congressional powers did not include forced testimony and therefore violations of a witness's rights guaranteed by the 1st, 4th, and 5th Amendments. This case was important because it set a legal precedent against the witch-hunt style of Congressional investigations in the McCarthy era.

This identification is quite good. The only important piece of data missing is the date, 1954.

3. The "Silent Majority"

Answer: During the 1972 election, Republican candidate Richard Nixon appealed to what he called the "silent majority" of Americans to support him. This "majority," he argued, was the core of "real" Americans who were religious, patriotic, and troubled by what Nixon saw as the unpatriotic and spoiled student unrest and Democratic party's pandering to the Left. He used this strategy to garner the support of those voters who voted for Wallace in 1968—conservatives who sought a return to the status quo. Along with his VP Agnew, who attacked the press and the students as rabble-rousers, Nixon won office in part on this appeal.

This identification is thorough and complete. Little space is wasted.

4. Philosophy of Martin Luther King

Answer: King, a Southern black Baptist minister, came to the forefront of the Civil Rights Movement in the mid-1950s with the Montgomery, AL bus boycott. He espoused a philosophy for challenging the white power structure and its violence against African Americans that combined elements of civil disobedience, evangelical Christianity, and non-violent resistance to authority. These three aspects formed the backbone of the protest strategy of the two main black Civil Rights bodies, the SCLC (King was president) and the SNCC (that is, until they became disillusioned with the gradualistic nature of this kind of protest in the late 1960s).

While the Civil Rights movement was a grass-roots coalition that included more than just King, he became the focus of media attention. His presence signaled both the important role of the black clergy and of determined non-violence (which appealed to whites more than Black Panther tactics) as a vehicle for civil rights change.

Good, but to make a better identification, the student should have included the sources for King's beliefs, such as Thoreau and Gandhi, as well as explained King's view of Christianity's emphasis on the redemptive aspects of suffering. Student also could have set King's philosophy in the larger history of African-American struggle for gradual vs. radical change (Booker T. Washington, W. E. B. DuBois, Marcus Garvey, etc.).

5. My Lai

Answer: My Lai is one of the gravest examples of U.S. Army atrocities in Vietnam. In March 1968, an Army unit invaded a small Vietnamese village under orders from its commander Calley. Soldiers killed hundreds of defenseless civilians, blew up and raped children, and burned the village to the ground. The argument was that in a guerilla war, where even children have been booby trapped, it was difficult to know who was the enemy. When information about this brutal event became public, domestic support for the war, already in crisis because of student protests, declined further, although U.S. support for the war continued embattled into Nixon's second term. My Lai raised the question of the role of personal consciences in warfare—or should a soldier follow orders even if they believe in their hearts that the order is morally wrong (a question put to German soldiers in WWII war crime trails)?

Although the language and syntax show signs of haste, the student covers the bases in this identification. He/she also situates the event of My Lai into a larger context of general warfare behavior, which elevates the quality of the identification.

Essay Questions: *Answer both questions for a total of 60 points (30 points each).*

1. In national mythology, the 1950s is characterized as a time of abundance, ease, and stability. Explore the accuracy of this stereotype.

ANSWER

When you think of the 1950s, you think of an easy-going President Eisenhower golfing, of clean-cut kids hanging out at a drive-in burger joint listening to the Everly Brothers, and of well-mowed lawns in new suburbs. Television, a booming new medium, reinforced this idea of the time in shows such as *I Love Lucy* and

The Donna Reed Show, where everyone was white, happy, and surrounded by lots of consumer products (including the TV itself). Americans who watched were told that "Father Knows Best"—in other words, don't worry or question what is going on around you, just leave it to the big guys.

This is a good start, in large part because it avoids vague abstraction. Notice how many specific details the student packs into the first paragraph.

Post-WWII America did in fact enjoy a booming economy, fueled by domestic and foreign consumption, and rising incomes (for many), as well as prominence as a superpower on the world stage. Yet, not all was apple pie and Chevrolet in the U.S. of the 1950s.

Beneath this surface of happiness, patriotism, and ease there existed (to borrow the title of Michael Harrington's 1962 study concluding that 1/4 of Americans lived in poverty), "The Other America."

Two brief paragraphs turn the picture of 1950s America around, signaling that the rest of the essay will deal with the darker side of the era.

Politically, the Cold War, in particular the Second Red Scare and our early covert involvement in Vietnam, provides the best challenge to the idea of America in the 1950s as a peaceful and prosperous time for all. The decade began with a war against Communist China and N. Korea, and ended with increased commitment to anti-Communism in Vietnam. Many Americans lived in constant fear of a Third World War, as evidenced by all the bomb shelters people built in their houses. The arms and space race with USSR further fueled these fears. Fear also fueled the Second Red Scare, during which two Americans, the Rosenbergs, were convicted and executed for being Soviet spies who sold information on how to build the bomb to the Russians. In an orchestrated crusade against Communism, Senator Joseph McCarthy of Wisconsin led the House Un-American Activities Committee to weed out all suspected Communists in this country. The HUAC quickly got out of control and turned into a witch-hunt, which attempted to deny Americans their Civil Rights (see above *Watkins v. U.S.* identification). The conformity of the 1950s clearly had a cost in terms of personal freedoms supposedly guaranteed by the U.S. Constitution.

The student begins with the strongest points he has to counter the image of an "apple pie and Chevrolet" America. If there were no time limit to this exam, it might be a better rhetorical strategy to work up dramatically to the strongest, most conclusive evidence. Usually, this structure makes for a more exciting and persuasive essay. However, when time is limited, you are best served by making your strongest points first—lest you run out of time before you can roll out your biggest guns.

Diplomatically, the United States' commitment to anti-Communism also got us involved in crises overseas. The CIA forced pro-U.S. governments in Iran, the Philippines, and Guatemala by coup, election control, and interventions that went against in many instances our alleged belief in popular sovereignty. The most blatant and long-reaching intervention came in Vietnam, where the U.S. sought strategic geographic control to stave off any Chinese expansion and to halt the domino theory effect (that one Communist country begets another and so on until "the map bleeds red"). Beginning under Truman, the U.S. government aided the French colonial government in Viet Nam against the rising poplar support for Ho Chi Minh. The French lost to popular forces in 1954, and the Geneva accords were in place, guaranteeing an election to unify the divided nation. Eisenhower ignored this international agreement and, with the CIA, put anti-communist leader Diem in charge of South Vietnam. Ironically, Ike's maneuver to halt Communism led to the establishment of the National Liberation Front and to increased American commitment, which resulted in a horrific and long war, which tore this country apart in the 1960s.

Note how the paragraph above, the one previous to it, and the next one begin with words that define their subject area: "politically," "diplomatically," and "socially." These leave no room for mistaking the subject of each paragraph.

Socially, America also witnessed strife in the 1950s that you would not know existed if you believed the decade's nostalgic stereotypes. A growing youth rebellion in the 1950s signaled some people's distaste for the dominant culture. The beatniks, for example, criticized white America's complacency and homogeneity. Even Elvis and his wagging hips and long hair signaled a shift from earlier milder forms of rock and roll. Unlike in the 1940s, when young and old listened to big band music, in the '50s rock and roll divided the generations. This alienation would only worsen with psychedelia in the 1960s. The most formidable challenge to the American status quo, however, came in the form of the black civil rights movement. This grass-roots movement had roots going back to slavery, but galvanized in 1954 with the Supreme Court's decision to overrule *Plessy v. Ferguson* (1896—separate but equal) with *Brown v. Board of Education.* This ruling declared segregation illegal and unconstitutional and gave struggling blacks legal leverage in their fight for equality. Many racist white Americans did not accept the ruling, and wanted to keep blacks down. During the Montgomery bus boycott, many white Americans beat and harassed blacks and even firebombed their homes, like that of movement leader Martin Luther King, Jr. This violence against blacks continued into the 1960s, with local cops turning fire hoses on marchers, governors refusing to allow black students to attend state-supported schools, and the resurgence of the KKK. America of the 1950s was not freedom and prosperity for all.

If you take as your definition of American a white middle-class family, then perhaps the stereotypes of the 1950s are accurate. However, this country is a melting pot, which includes a wide variety of races, creeds, ideologies, and incomes and cannot be so stereotyped without sacrificing a more accurate sense of America's complicated experience in the 1950s. You could argue that the complacence and ease of some Americans in the 1950s meant intolerance and poverty for others, and you would be right. There is always another side to the story. As students of history, we must always try to remember the "other America" so our picture of the past can try to be as complete as possible.

For the most part, this is a very well-written essay, which illustrates a fine command of the data discussed in class concerning the 1950s. It is organized into clear-cut paragraphs, all of which concern a central idea or theme. Instead of merely cataloging the "truth versus myth" of the 1950s, as many students are wont to do, this student actually makes a cogent argument about larger issues, such as the need to be sensitive to the variety of ways different groups experience the past, based upon their race, generation, class, etc. The student also exhibits an attuned awareness of critical thought and of owning historical material to persuade the reader to one's central claims. Well done.

2. **Many conversations have taken place in the news media and perhaps around American dinner tables about the limits of executive privilege. In this course, we have come across several moments in American history when various elements of society and government have challenged the rights and powers of the executive branch. Choosing two or three such events, explain your position on the question of executive privilege.**

ANSWER

When the United States was born out of the American Revolution, its new citizens created a form of government that differed significantly from the British monarchical system of which they had been a long-suffering part. While some new Americans wanted to make their president a king, the majority, including Washington himself, resisted and instead the character of the presidency as a human institution was established. The powers of the president were not divinely guaranteed, but rather they were shaped over time by external forces, including Congress, the political parties, and American popular opinion. Three moments in U.S. history from 1865 best illustrate this process of debating the limits of presidential powers: Andrew Johnson versus Congress during early Reconstruction; Franklin Roosevelt's court-packing controversy; and Richard Nixon's resignation after Watergate. All show how important the American government's inherent system of checks and balances is and how no one person, in the form of the president, should be above the democratic laws of the nation.

An excellent opening paragraph. It shows a broad grasp of the historical context of presidential power and executive privilege, then it clearly announces and defines its focus, being careful to list the events that will be used to illustrate the points made. Beginning this way also demonstrates a complete understanding of the terms of the question. This signals to the instructor that she is about to read a thoughtful and competent answer. It also sets up the structure of the rest of the essay—making it that much easier to write (and to read).

When the Civil War ended, the government struggled over how to readmit the former Confederate states into the union. President Lincoln proposed a rather easygoing plan, which would have readmitted states with 10 percent of the population willing to take loyalty oaths to the union, and thus vote in loyal state governments. Certain people, like ex-Confederate officers and government officials, would have been denied the vote. But in April 1865, Lincoln was assassinated, and Vice President Andrew Johnson became president. The Republican-controlled Congress, split between moderates and Radicals, each of which had different ideas of what Reconstruction should look like, thought Johnson, a long-time enemy of the antebellum South's planter elite, would propose a harsher plan. Instead, his Presidential Reconstruction was very lenient in their eyes.

Johnson enacted his plan, which included the return of most property (except slaves) and liberal pardons for rebels, when Congress was out of session and angered Republican leaders. The result was southern state governments run by many of the same people who had rebelled in the first place. For example, Georgia elected Alexander Stephens, who had been VP of the Confederacy, to the U.S. Senate! And these states also enacted laws called "black codes," which limited freed people's movements and rights (such as intermarriage, testimony in court against whites, and agricultural contracts which reinforced working conditions on plantations that were similar to slavery). Radicals and moderates in Congress were angry with Johnson's leniency and, despite Johnson's vetoes, passed the Freedmen's Bureau and Civil Rights acts to counteract his presidential plan. Republicans and Johnson were alienated from each other over the extent of presidential versus Congressional power. This came to a head in 1867 when Johnson suspended Secretary of War Stanton, a Radical, again while Congress was out of session, and thus ignored the recently passed Tenure of Office Act, which said cabinet members had to serve the entire term of the president who appointed them, unless fired by the Senate only. Congress tried Johnson for impeachment and for overstepping presidential powers by ignoring the work of the legislative branch. In the end, Johnson was found not guilty, but his future as a president in the next election was over. By overstepping his office's boundaries, Johnson was effectively removed from being a challenge to the more Radical Reconstruction that ensued.

A rather one-sided viewpoint against Andrew Johnson, but a defensible one—and, more important, a reasonably well-argued viewpoint that supports the student's thesis. Few professors will split hairs in interpreting a time-limited essay response. After all, entire volumes have been devoted to the controversial and hapless Andrew Johnson.

Seventy years later, after winning his second election, Franklin Roosevelt also overstepped his powers as president, and a crisis ensued, this time with another branch of government, the judiciary. In an attempt to "pack the court" with pro-New Deal judges and to outnumber the older, arch-conservative judges who voted against the constitutionality of FDR's programs like the National Recovery Act and who threatened to do the same to the Social Security Act, which FDR fought for on the campaign trail, Roosevelt proposed court reform. He wanted the president to have the power to appoint a new judge for every existing judge over the age of seventy, up to a total of six new judges. FDR claimed this reform was to lighten the load for older judges, when in fact he hoped to influence votes. The government and the public were outraged at his blatant political manipulation with the process, despite his otherwise strong popularity. FDR's move was seen by many as a grab for total power, not unlike that of Johnson. This kind of total autocratic power goes against American ideals and backfired on Roosevelt, and his reform plan was defeated. And Roosevelt's image was stained.

A more recent example of the limits of executive power is the Watergate scandal, which led to President Richard Nixon's disgraced resignation and departure from office in August 1974. During the 1972 presidential campaign, several burglars broke into Democratic party offices in the Watergate Hotel in D.C. Federal judge John Sirica refused to believe that they acted on their own and began an investigation into connections between them and the White House. One burglar, McCord, confessed that high-up aides of the president knew in advance of the break-in and had urged him and his fellow thieves to remain silent about the connection. As the Senate's investigation edged closer to the Oval Office, Nixon fired his special counsel, John Dean, and forced the resignation of others, arguing publically that he was on the side of getting to the bottom of the crisis. He also oversaw the appointment of a special prosecutor, Cox. The investigation unearthed a paranoid White House, with a presidential enemy list, illegal campaign donations, and use by Nixon of government agencies to spy on his enemies, as well as the very active involvement of the White House in the coverup of the break-in (on John Dean's turncoat testimony v. Nixon). When the existence of Oval Office tapes became known, the Senate committee demanded them, and Nixon refused, citing presidential privilege and fears about national security. Cox ordered the release of the tapes, and Nixon had him fired, and impeachment proceedings began. Nixon released some edited tapes, which revealed him to be a petty, paranoid, power-hungry man, and his public image suffered. He resigned in disgrace, to end the impeachment process and the court-ordered release of more tapes, in August 1974.

This event not only forever changed the untouchable status of presidents in this country and set a precedent for a no-holds-barred media, but it also clearly declared that the president was not above the law and that presidential privilege was not ironclad.

The three instances of the abuse of men of the office of president illustrate why we should continue to have a three-branched system of checks and balances in this country to counterbalance human tendencies toward power-hungry, undemocratic behavior in office. If the president is above the people, who will speak and act for the people? He or she must be subject to the same criteria as the rest of us.

Another well-written essay, exhibiting the student's command of the material. It is organized into three clear and discrete examples. This essay tends to be more narrative than argumentative, despite the question's explicit request for the student to argue his/her viewpoint. Given the examination context in which this was written, however, the very fact that the student has a thesis statement about his/her views on the limits of presidential powers, which is then reiterated in the conclusion, serves to frame the data well. It would be improved by interjecting the student's views on presidential powers in relation to the specific examples discussed throughout the essay.

PART FOUR

FOR YOUR REFERENCE

CHAPTER 26

CHRONOLOGY OF THE MAJOR EVENTS IN U.S. HISTORY

1492 Columbus makes his first voyage.

1497 Englishman John Cabot, sailing for Italy, explores North America from Canada to Delaware.

1513 Juan Ponce de León explores Florida.

1524 Giovanni da Verrazano leads French expedition along the coast from Carolina to Nova Scotia.

1565 St. Augustine, Florida, founded.

1607 Jamestown, Virginia, founded: first permanent English settlement in America.

1609 Henry Hudson (sailing for the Dutch) explores New York harbor and the Hudson River. Spain settles Santa Fe, New Mexico.

1619 First black slaves sold in Virginia.

1620 Pilgrims land in Plymouth, Massachusetts; Mayflower Compact concluded.

1623 Dutch found New Netherland (later New York).

1626 Peter Minuit buys Manhattan Island from Native Americans.

1630 The Massachusetts Bay Colony founded.

1634 Maryland founded.

1636 Harvard, the first college in America, is founded. Roger Williams, exiled from Massachusetts, founds Providence, Rhode Island.

1664 The English capture New Netherland; becomes New York.

1682 William Penn founds Pennsylvania.

1688 Pennsylvania Quakers make the first formal protest against slavery.

1692 Salem, Massachusetts, witchcraft trials and executions.

1732 Georgia founded.

1754 French and Indian War begins.

1763 The French and Indian War ends.

1764 Sugar Act passed.

1765 Stamp Act passed: incites colonial opposition to taxation without representation.

1766 Repeal of Stamp Act.

1767 Townshend Acts passed.

1770 Boston Massacre

1773 Boston Tea Party

1774 Intolerable Acts passed.

1775 American Revolution begins with the battles of Lexington and Concord.

1776 Declaration of Independence drafted and signed.

1777 Washington defeats Cornwallis at the battle of Princeton; Burgoyne captures Fort Ticonderoga, but is defeated at Saratoga.

1778 France allies with the United States.

1781 U.S.-French victory at Yorktown, last major battle of the Revolutionary War.

1783 Treaty of Paris ends the Revolutionary War.

1787 Constitutional Convention convenes.

1789 Washington elected. John Adams is vice president; Thomas Jefferson, secretary of state; Alexander Hamilton, secretary of the treasury.

1791 The Bill of Rights enacted.

1793 Cotton gin invented by Eli Whitney; revives slavery in the South.

1794 Whiskey Rebellion

1801 Tripolitan War begins.

1803 *Marbury v. Madison* Supreme Court decision. Louisiana Purchase made.

1804 Lewis and Clark expedition begins. Burr-Hamilton duel.

1805 Tripolitan War ends.

1808 Importation of slaves outlawed.

1812 War of 1812 begins.

1814 War of 1812 ends (Treaty of Ghent).

1820 Missouri Compromise

1825 Erie Canal opened.

1830 Indian Removal Act

1831 Nat Turner slave rebellion in Virginia.

1836 The Alamo besieged and captured; Texas declares independence from Mexico.

1844 Inventor Samuel F. B. Morse sends first telegraph message (from Washington to Baltimore).

1846 U.S.-Mexican War begins.

1848 U.S.-Mexican War ends (Treaty of Guadalupe Hidalgo). Gold is discovered in California.

1849 Gold Rush begins.

1850 Compromise of 1850.

1852 Harriet Beecher Stowe's *Uncle Tom's Cabin* published.

1854 Republican Party formed.

1857 Dred Scott decision.

1858 Lincoln-Douglas debates.

1859 John Brown raids Harpers Ferry.

1861 Civil War begins.

1862 Homestead Act.

1863 Gettysburg Address and Emancipation Proclamation.

1864 Sherman marches through Georgia; Atlanta burned.

1865 Civil War ends. President Lincoln is assassinated. Thirteenth Amendment abolishes slavery.

1868 President Johnson impeached, but acquitted.

1869 Transcontinental railroad completed.

1870 Great Chicago Fire.

1872 Susan B. Anthony arrested for voting. Amnesty Act restores rights to Southern citizens. Yellowstone becomes first U.S. national park.

1875 Civil Rights Act.

1876 Custer defeated at the battle of the Little Bighorn.

1877 United States violates its treaty with the Dakota Sioux by seizing the Black Hills.

1881 Sitting Bull surrenders. President Garfield assassinated. Booker T. Washington founds Tuskegee Institute.

1883 Civil Rights Act of 1875 is partially invalidated by the Supreme Court.

1886 Haymarket Square riot in Chicago. Geronimo surrenders.

1890 Wounded Knee massacre ends the Indian Wars. Ellis Island begins operation as a principal port of entry for immigrants.

1898 Spanish-American War.

1899 Open Door Policy: opens China to trade, but respects its sovereignty as a nation.

1901 President McKinley assassinated.

1903 Panama Canal Treaty. Orville and Wilbur Wright make the first airplane flight.

1906 San Francisco earthquake.

1908 The United States bars Japanese immigration.

1911 Triangle Shirt Waist Company fire (New York City) provokes demands for reforms in working conditions.

1913 Sixteenth Amendment authorizes income tax.

1914 Colorado National Guard burns a striking miners' camp in the Ludlow Massacre.

1916 U.S. Marines occupy the Dominican Republic.

1917 Puerto Rico becomes a U.S. territory. United States enters World War I.

1918 World War I ends.

1919 Congress overrides President Wilson's veto of Prohibition.

1920 Red Scare and "Palmer raids." Prohibition begins.

1921 Immigration quotas set.

1925 Scopes "Monkey trial" in Tennessee.

1927 Charles Lindbergh makes the first trans-Atlantic solo flight.

1929 Stock market crashes, beginning the Great Depression.

1933 President Franklin Roosevelt's "100 days": massive federal legislation to attack the Depression.

1935 The Works Projects Administration (WPA) established; Social Security Act signed.

1939 World War II begins in Europe.

1940 First U.S. peacetime draft enacted.

1941 Japanese attack Pearl Harbor, bringing U.S. into World War II.

1942 120,000 Japanese-Americans interned on the West Coast.

1943 Racial discrimination outlawed among war contractors.

1944 D-Day landing at Normandy.

1945 FDR dies; Truman becomes President. Nazi Germany surrenders. Atomic bombs dropped on Japan, which surrenders. United Nations Charter adopted.

1946 Philippines given independence.

1947 Cold War begins. Marshall Plan announced. Central Intelligence Agency (CIA) and the National Security Council established under the National Security Act. New Red Scare: Hollywood figures investigated.

1949 North Atlantic Treaty Organization (NATO) formed.

1950 Senator Joseph McCarthy accuses State Department employees of Communist party affiliations. Korean War begins.

1951 Julius and Ethel Rosenberg convicted of espionage.

1952 U.S. tests the world's first hydrogen bomb. Immigration and Naturalization Act lifts racial and ethnic barriers to naturalization.

1953 Rosenbergs executed. Korean War ends in ceasefire.

1954 Supreme Court rules segregated public schools unconstitutional (*Brown v. Board of Education of Topeka*). Senate censures Joseph McCarthy. Southeast Asia Treaty Organization (SEATO) formed.

1955 Rosa Parks arrested for failing to relinquish her bus seat to a white passenger: birth of the modern civil rights movement. U.S. military advisers sent to South Vietnam.

1956 Federal Aid Highway Act: interstate highway system construction begun.

1957 Little Rock Central High School desegregation crisis resolved by federal troops.

1958 First U.S. satellite launched into orbit.

1959 Alaska and Hawaii become the forty-ninth and fiftieth states.

1960 Sit-ins in Greensboro, North Carolina, protest cafeteria's refusal to serve African Americans.

1961 U.S. severs diplomatic relations with Cuba.
Bay of Pigs invasion fails. Alan B. Shepard, Jr., is first U.S. man in space.

1962 Cuban Missile Crisis. John H. Glenn, Jr., becomes first American to orbit in space.

1963 Martin Luther King. Jr. leads civil rights march on Washington, D.C. White House-Kremlin "hot line" installed. JFK assassinated.

1964 Twenty-fourth Amendment outlaws the poll tax in federal elections. Civil Rights Act passed by Congress. Gulf of Tonkin Resolution gives President Lyndon Johnson broad power to wage war in Vietnam and Indochina.

1966 Medicare begins.

1967 Thurgood Marshall becomes the first black Supreme Court justice. Antiwar protests and race-related rioting sweep the nation.

1968 My Lai massacre in Vietnam. Martin Luther King, Jr. assassinated. Presidential candidate Robert F. Kennedy assassinated.

1969 "Stonewall rebellion" is the start of the modern gay rights movement. Woodstock festival. U.S. begins peace talks with Vietnam, as troop withdrawal starts. Neil Armstrong is the first man to walk on the moon.

1970 Kent State massacre. Environmental Protection Agency (EPA) is established.

1971 Massive antiwar demonstration in Washington, D.C.

1972 Watergate break-in. Nixon visits China.

1973 Paris Peace Accords end Vietnam War. Spiro T. Agnew resigns as vice president, and Gerald R. Ford is appointed to replace him. Congress overrides Nixon's veto of the War Powers Act, effectively cutting off funding for the Vietnam War.

1974 House Judiciary Committee votes Articles of Impeachment against President Nixon, who resigns. President Ford pardons Nixon.

1975 North Vietnamese troops enter Saigon.

1976 U.S. celebrates its Bicentennial.

1977 President Jimmy Carter grants amnesty to Vietnam draft resisters. National Women's Conference convenes in Houston.

1978 Senate votes to return the Panama Canal to Panama. President Carter brokers a peace between Egypt and Israel.

1979 The Three Mile Island nuclear accident occurs. Iran takes U.S. embassy personnel hostage.

1981 Iran hostages released. Attempt made on President Reagan's life.

1982 The Equal Rights Amendment (ERA) fails to be ratified.

1983 U.S. Marines and Army Rangers invade the island of Grenada; applying the War Powers Act, Congress demands that the troops leave Grenada.

1984 U.S. imposes economic sanctions on South Africa to protest apartheid and other human rights violations. Major U.S.-Soviet summit meeting as relations thaw.

1985 U.S. and Soviet Union resume arms-reduction talks. Mikhail Gorbachev becomes Soviet premier; U.S.-Soviet relations continue to improve.

1986 Space shuttle Challenger explodes. U.S. bombs Tripoli and Benghazi, Libya, in retaliation against terrorism.

1987 Iran-contra scandal. Wall Street scandals. "Black Monday" stock market crash.

1989-91 Cold War ends. Persian Gulf War (Desert Storm).

1991 Los Angeles race riots: worst in American history.

1992 H. Ross Perot runs for president: strongest third-party showing in U.S. history.

1995 Domestic terrorism: federal building in Oklahoma City bombed.

1996 American economy resurgent; Bill Clinton elected to a second term.

1998 Clinton sex and perjury scandal; impeachment inquiry begun.

1999 Impeachment trial and aquittal of President Clinton.

CHAPTER 27

RECOMMENDED READING

Surveys

John A. Garraty and Robert A. McCaughey, *The American Nation*, 6th Ed. (1987)

Samuel Eliot Morison, Henry Steele Commager, and William E. Leuchtenburg, *The Growth of the American Republic*, 7th Ed. (1980)

Richard B. Morris, ed., *Encyclopaedia of American History*, 6th ed. (1982)

Age of Discovery

Samuel Eliot Morison, *The European Discovery of America*, 2 Vol. (1971-74)

David B. Quinn, *North America from Earliest Discovery to First Settlements: The Norse Voyages to 1612* (1977)

The Colonial Era

Charles M. Andrews, *the Colonial Period of American History*, 4 Vol. (1934-38; reprint ed., 1964)

James Axtell, *The European and the Indian* (1982)

Daniel J. Boorstin, *The Americans: The Colonial Experience* (1958; Reprint ed., 1988)

Lawrence Henry Gipson, *The British Empire Before the American Revolution*, 15 vols. (1936-70)

Alan Heimert, *Religion and the American Mind: From the Great Awakening to the Revolution* (1966)

Francis Jennings, *The Invasion of America* (1975)

Henry F. May, *The Enlightenment in America* (1976)

Gary B. Nash, *Red, White, and Black: The Peoples of Early America*, 2nd Ed. (1982)

Howard H. Peckham, *The Colonial Wars, 1689-1762* (1964)

The Revolution

John R. Alden, *The American Revolution, 1775-1783* (1954; reprint ed., 1962)

Bernard Bailyn, *The Ideological Origins of the American Revolution* (1967)

Edward Countryman, *The American Revolution* (1985)

Don Higginbotham, *The War of American Independence* (1971; reprint ed., 1985)

Charles Royster, *A Revolutionary People at War* (1979; reprint ed., 1981)

Federal Period

Noble E. Cunningham, *The Jeffersonian Republicans* (1957)

Richard Hofstadter, *The Idea of a Party System* (1969)

Gerald Stourzh, *Alexander Hamilton and the Idea of Republican Government* (1970)

Gordon S. Wood, *The Creation of the American Republic, 1776-1787* (1969)

Antebellum Period

K. Jack Bauer, *The Mexican War, 1846-1848* (1974)

George Dangerfield, *The Era of Good Feelings* (1952, reprint ed., 1973)

Bernard De Voto, *The Year of Decision, 1846* (1942, reprint ed., 1989)

Eugene D. Genovese, *Roll, Jordan, Roll: The World the Slaves Made* (1974)

Glover Moore, *The Missouri Controversy, 1819-1821* (1953; reprint ed., 1967)

Michael Paul Rogin, *Fathers and Children: Andrew Jackson and the Subjugation of the American Indian* (1975)

Arthur M. Schlesinger, Jr., *The Age of Jackson* (1945; reprint ed., 1953)

Alexis De Tocqueville, *Democracy in America*, 2 vols. (1835; many later editions)

The Civil War

James M. McPherson, *Ordeal by Fire* (1982) and *Battle Cry of Freedom* (1988)

J. G. Randall and David Donald, *The Civil War and Reconstruction*, 2nd ed., rev. (1969)

Allan Nevins, *Ordeal of the Union*, 8 vols. (1947-71)

Kenneth M. Stampp, *The Peculiar Institution* (1956, reprint ed., 1978)

Reconstruction

Eric Foner, *Reconstruction: America's Unfinished Revolution*, 1863-1877 (1988)

John Hope Franklin, *Reconstruction* (1961)

Leon F. Litwack, *Been in the Storm So Long: The Aftermath of Slavery* (1979)

Rembert W. Patrick, *The Reconstruction of the Nation* (1967)

Kenneth M. Stampp, *The Era of Reconstruction, 1865-1877* (1965; reprint ed., 1975)

C. Vann Woodward, *Reunion and Reaction* (1951; reprint ed., 1966) and *the Strange Career of Jim Crow*, 3rd Rev. Ed. (1974; reprint ed., 1982)

Era of National Expansion

Ray Allen Billington and Martin Ridge, *Westward Expansion*, 5th Ed. (1982)

Harold E. Briggs, *Frontiers of the Northwest: A History of the Upper Missouri Valley* (1940; reprint ed., 1950)

Sean Dennis Cashman, *America in the Gilded Age: From the Death of Lincoln to the Rise of Theodore Roosevelt*, 2nd Ed. (1988)

Harold U. Faulkner, *Politics, Reform, and Expansion, 1890-1900* (1959; reprint ed., 1963)

Lawrence Goodwyn, *Democratic Promise: The Populist Moment in America* (1976)

Samuel P. Hays, *The Response to Industrialism, 1885-1914* (1957)

John D. Hicks, *The Populist Revolt* (1931; reprint ed., 1981)

Edward C. Kirkland, *Industry Comes of Age* (1961)

Rodman W. Paul, *The Far West and the Great Plains in Transition, 1859-1900* (1988)

Norman J. Ware, *The Labor Movement in the United States, 1860-1895* (1929; reprint ed., 1964).

Walter Prescott Webb, *The Great Plains* (1931; reprint ed., 1981)

Leonard D. White, *The Republican Era, 1869-1901* (1958; reprint ed., 1965)

Imperialism and Progressivism

Robert H. Ferrell, *Woodrow Wilson and World War I, 1917-1921* (1985)

A. Whitney Griswold, *The Far Eastern Policy of the United States* (1938; reprint ed., 1966)

Walter Lafeber, *The New Empire: An Interpretation of American Expansion, 1860-1898* (1963)

Arthur S. Link, *Woodrow Wilson and the Progressive Era, 1910-1917* (1954; reprint ed., 1963)

Ernest R. May, *Imperial Democracy* (1961; reprint ed., 1973) and *The World War and American Isolation, 1914-1917* (1959)

Richard L. McCormick, *Progressivism* (1983)

George E. Mowry, *The Era of Theodore Roosevelt, 1900-1912 (1958; reprint ed., 1962)*

Dana G. Munro, *Intervention and Dollar Diplomacy in the Caribbean, 1900-1921* (1964; reprint ed., 1980).

David F. Trask, *The War with Spain* (1981)

Neil A. Wynn, *From Progressivism to Prosperity: World War I and American Society* (1986)

Interwar Period and World War II

Irving Bernstein, *Turbulent Years: A History of the American Worker, 1933-1941* (1969)

John Morton Blum, *V Was for Victory: Politics and American Culture During World War II* (1976)

Norman H. Clark, *Deliver Us from Evil* (1976) (Prohibition)

Kenneth S. Davis, *Experience of War: The United States in World War II* (1965)

William E. Leuchtenburg, *The Perils of Prosperity, 1914-32* (1958) and *Franklin D. Roosevelt and the New Deal, 1932-1940* (1963)

Geoffrey Perrett, *America in the Twenties* (1982)

Postwar to the Present

Taylor Branch, *Parting the Waters: America in the King Years, 1954-1963* (1988)

John Lewis Gaddis, *Strategies of Containment* (1982)

George C. Herring, *America's Longest War: The United States and Vietnam, 1950-1975*, 2nd Ed. (1986)

Burton I. Kaufman, *The Korean War* (1986)

Ralph B. Levering, *The Cold War, 1945-1987*, 2nd Ed. (1988)

William L. O'Neill, *Coming Apart: An Informal History of America in the 1960's* (1971)

Frederick F. Siegel, *Troubled Journey: From Pearl Harbor to Ronald Reagan* (1984)

Note: This index covers Part One: Preparing Yourself and Part Two: Study Guide, pages vii-127 of the text. Material in the sample exams, pages 130-318, is not indexed.